Institutional innovation in water management: the Scottish experience

by

W. R. D. Sewell, J. T. Coppock

and Alan Pitkethly

GEO·BOOKS

British Library Cataloguing in Publication Data

Sewell, W.R. Derrick
Institutional innovation in water management:
the Scottish experience.
1. Water resources development
I. Title II. Coppock, J.T. III. Pitkethly, Alan
333.91 TC409

ISBN 0 86094 189 2

Published by Geo Books
Regency House
34 Duke St.
Norwich NR3 3AP
UK

Printed by Short Run Press Ltd., Exeter, Devon

Contents

Chapter

List of figures

List of tables

Preface

There is a great deal of discussion of water problems these days. Population expansion and industrialization have brought major problems of water shortages and water pollution to both developing and the developed countries of the world. In some regions, notably the Sahel of Africa, droughts have caused both losses of life and starvation on a massive scale. Floods continue to wreak havoc with property and result in huge losses of income, and in many countries, losses of life as well. Understandably, therefore, water problems have become a priority item on the political agenda of several countries. And at the global level, the United Nations, the World Bank, and UNESCO have identified water development as a key element in the solution of the world's food, poverty, and health problems.

The major hope for success lies in innovation, in part, on the technological front but especially with respect to institutions, and particularly the laws, policies and administrative arrangements that are concerned with water managment. Often the needed technology is available but existing institutions may prevent or fail to encourage its adoption. Moreover, the introduction of technology may not always be needed. In some cases a shift in human behaviour may require no new facilities whatsoever, as would be the case, for example, if pricing systems encouraged water conservation, or if land use regulations discouraged damage-prone activities from occupying flood plains.

During the past two or three decades there has been a good deal of experiment with water institutions, particularly in Western Europe and North America. Several different countries facing broadly similar water management problems have taken quite different routes in the attempt to solve them. At the same time, while some regions within individual countries have shown a high propensity to innovate, others have experienced little or no change.

The reasons are not always to be found in the absence of problems. On the contrary, the greatest conservatism has often been in regions with the most severe difficulties.

Thus far there has been little inquiry into the nature and pace of institutional innovation in water management. Yet it is clear that research on this matter is urgently needed. For instance, are there essential differences between such innovation and that in other aspects of human activity, such as road development, housing, or manufacturing? Moreover, given that modification of existing laws, policies, or administrative arrangements may be costly in financial and other terms, what seem to be the ingredients for success of the various innovations that have been tried? Obviously, the answers to such questions would be of considerable interest not only to academics, but also to water managers, politicians and others.

The authors of this volume, cognizant of the need for insights into institutional innovation in water management, began research on the subject in the mid-1970s. Stimulated by the facts that some major changes to the approach to such management had been undertaken in England and Wales in 1963 and 1973, as well as in France and North America, that progress in implementing the new properties had varied considerably from region to region, and that there appeared to be important differences with the way things were done in Scotland, they embarked upon an in-depth analysis of the experience in the latter. The basic aim is to determine the extent to which a number of concepts of contemporary water management had been adopted in Scotland and to account for variations in their acceptance. The roots of the present management system were traced back to the 1930s, when the major antecedents of the present system seem to have appeared. The aim was to identify various economic, social and political influences in this evolution. Reliance was placed upon official documents as well as the writings of academics and other observers. In addition, the authors interviewed numerous individuals who had been involved in the development of Scotland's water management institutions, officials of water agencies, technical advisors, politicians and members of various interest groups. The perceptions of the problems to be faced as well as applicable solutions varied considerably among the different participants. Knowledge of these viewpoints as well as the roles of the latter helps shed important light on the nature and pace of institutional change.

The study was a joint venture, organized by a researcher from Canada, Professor Derrick Sewell, and one from Scotland, Professor J. T. Coppock, with the collaboration of a Scot who now works in England. All three were familiar with developments in England and Wales. In addition, Professor Sewell had done a good deal of research in other countries including those in Western Europe and North America They were aided in their endeavours by research councils and by various agencies, officials and researchers on both sides of the Atlantic.

The authors would like to acknowledge, in particular the financial assistance provided by the Canada Council, the Social Sciences and Humanities Research Council of Canada, the National Advisory Committee on Water Resources Research of Canada, and the Social Science Research Council in the United Kingdom. They are also extremely grateful for the help provided by the officials of water management agencies and other bodies in Scotland, at both central government and local government levels of administration, as well as others in the private sector and at universities. Space and agreements to maintain confidentiality do not permit a full listing of all those who participated. Special mention, however, must be made of the advice provided by the following people: J. W. Shiell, R. H. Cuthbertson, J. I. Waddington, D. Hammerton, D. Kinnersley.

In addition, the authors were assisted by several individuals who undertook library research or who carried out interviews at various stages of the investigation, notably Sandra Laing, Carole Dixon, Jill Dunn, Susan Cartwright, Lorna Barr, and John Newcomb.

Finally, the authors wish to acknowledge the patient efforts of those who typed various drafts of the manuscripts and those who prepared the illustrations. Thanks are due particularly to: Elizabeth McDougall, Eleanor Lowther, Gillian Levy, Debbie Penner, Jennifer Hobson-Roy, Ian Norie, Ole Heggen, Ken Josephson, and Diane Brazier.

W. R. Derrick Sewell,
J. T. Coppock
Alan Pitkethly

Edinburgh, November 1983

1

Institutional innovation in water management

The development of more effective institutions of water management has become a major challenge for planners and policy-makers in Europe and North America. Despite the advances that have been made in technology and the huge expenditures that have been incurred in developing water supply, systems of waste disposal and other water services, there remain major problems, associated particularly with scarcity of supplies, pollution, flooding and conflicts in the use of water. Convinced that the remedy to these deficiencies lies in the improvement of policies and administrative frameworks, governments in several countries have undertaken radical modifications of such frameworks in the past two decades. Elsewhere, less radical solutions have been adopted. The aim of this book is to examine what happened in Scotland in the postwar period, with particular reference to supply, sewerage and water quality, and to consider how far Scottish experience resembles that in other countries. It is based on documentary sources and on extensive discussions with those responsible at various levels for Scottish water management, though the confidential nature of the latter means that they have provided insights rather than direct evidence.

THE MOVE TOWARDS REFORM

There are numerous illustrations of the move towards reform. In 1963, for example, the British Parliament passed a Water Resources Act, calling for the establishment of 29 River Authorities covering the entire area of England and Wales (Parker and Penning-Rowsell, 1980). The aim was to foster a more integrated approach to water management, particularly with respect to water supply, pollution control, land drainage and fisheries. The authorities were to establish hydrometric networks, prepare comprehensive, long-term plans and introduce a charging scheme for water abstraction. At the same time, steps were being taken to improve the quality of rivers through the licensing of discharges and the more rigorous application of fines. These River Authorities were replaced in 1973 by nine Regional Water Authorities in England and a Welsh Water Development Authority, with comprehensive responsibilities for all aspects of the water cycle.

In 1964 the French Government established six river basin agencies, called Agences Financieres de Bassin (Nicolazo-Crach and Le Frou, 1978), to develop a more effective approach to water supply and facilitate a sustained attack on water pollution. Specifically, the agencies were to prepare water management plans which would be revised periodically as new problems and new concepts emerged. In addition, they were to impose charges not only for the abstraction of water but also for the disposal of effluents into rivers. The revenue derived from these fees was to be used to assist industries and municipalities in constructing new facilities for water supply and water purification.

In Canada the Federal Parliament passed a new Water Act in 1970, giving the government powers to enter into co-operative agreements with provincial authorities to draft long-range plans for the development of river basins and to instituted firm measures for the control of water pollution (Foster and Sewell, 1981). Several agreements have been made under the Act and tougher legislation has been introduced to deal with water pollution (Woodrow, 1980). A Department of the Environment was established, bringing together several agencies concerned with resources and environmental management. The various provinces have also introduced new legislation to deal with water problems and have made major modifications to administrative structures, aiming particularly to improve co-ordination of functions (Tate, 1981; Mitchell and Gardner, 1983).

In the United States a National Water Commission was formed in 1968 to provide a broad assessment of the major water problems facing the United States and to prepare guidelines for the kinds of action that might be taken to deal with them (U.S. National Water Commission, 1973). Its report, *Water Policies for the Future* published in 1973, constitutes one of the most comprehensive reviews of water issues ever undertaken in North America and its recommendations represent a bold attempt to deal not only with the existing problems but also with those which will become acute by the end of the century. Meanwhile, several important modifications have been made to legislation, policies and administrative structures relating to water management in the United States, particularly with respect to supply, pollution and floods. For example, experiments have been conducted to find better ways of co-ordinating activities within and between various levels of administration (OECD, 1972).

There have been major alterations to water institutions in other countries too in the past two decades, especially in Western Europe (Johnson and Brown, 1976; OECD, 1976). The responses, like those already described, have varied from one country to another, reflecting differences in culture, traditions and political philosophies, as well as differences in the problems encountered. There are, however, some common threads. Water management is becoming increasingly complex and existing institutions seem unable to cope. The costs of water development are soaring, and water agencies have to compete for scarce public funds with an increasing range of other services. At the same time there is pressure for environmental conservation, which limits the range of opportunities that can be considered for development. The challenge is to find ways of improving efficiency, while at the same time maintaining the standard of service and ensuring the support of those who pay for it.

Where change has occurred it has typically been incremental and *ad hoc*. Sometimes it has consisted of little more than the addition of a new clause to an existing law or a new name for an agency. Often the modifications are responses to crises rather than the result of rational deliberations which place the problem in the broadest possible context. In many instances, it can be fairly said, first-aid has been applied where major surgery would have been more appropriate. Nevertheless, as the preceding examples have shown, bold attempts have been made in some countries to improve the institutional framework by introducing a different approach to the one that existed before. This was certainly the case with the 1973 Water Act in England and Wales, the 1964 Water Law in France and the 1970 Water Act in Canada. Application of the philosophy underlying the National Water Commission's report would likewise lead to major changes in water institutions in the United States.

THEMES IN CONTEMPORARY MANAGEMENT

The various institutional adjustments that have taken place embrace one or more of a number of themes that have appeared in contemporary water management in Europe and North America. Such themes reflect not only the problems that are being faced but also the concepts that are being brought to bear in the search for solutions. They also provide a yardstick against which the institutions in any particular region or country can be assessed, given similarities of problems, stages of economic development, political philosophy and so on. The most sophisticated contemporary frameworks are characterized by:

(a) a broadening of the perspective of water management;

(b) integration of purposes and consolidation of functions within and between various levels of administration;

(c) consideration of an expanding range of strategies in the search for solutions to water problems;

(d) recognition of water as an economic good;

(e) provision of greater opportunities for public inputs into planning and policy-making.

A Broadening Perspective

The past half century has witnessed a marked broadening of the perspective of water management. At the turn of the 20th century it was characterized by a concern for local problems, with water being considered in isolation from other resources and little attention being paid to the environmental, economic or social impacts of development. This perspective prevails in many parts of the world even today. But, as White (1969) has shown, the emergence of new problems, improved technologies and new ideas, coupled with shifts in value systems, has led to a substantial broadening of the approach in North America (Figure 1.1). Experience in much of Western Europe, Australia, New Zealand and a number of countries in the Third World has been similar. There has been a progressive shift from single-purpose construction to multiple-purpose construction, followed in some cases by a shift to water management by multiple means, and in a few instances, the integration of such management with plans for overall economic and social development.

In some countries this change in perspective has been formally recognized in the institutional framework (United Nations, 1973, 1974). In France, for example, the Basin agencies undertake their planning within the constraints of the government's overall five-year plan and the Directors of the Agencies have close communication with the Commissariat du Plan. In Hungary, co-ordination between water planning and economic planning is accomplished through the President of the National Water Authority being a member of the National Economic Planning Council. In Israel, the Ministerial Committee of Economic Affairs co-ordinates policy decisions in the water field with those concerning economic and social development.

Integration

A second feature of contemporary water management is the trend towards integration of purposes and consolidation of functions. Water is managed for a wide variety of purposes, including water supply, waste disposal, generation of hydroelectricity, irrigation, water-based recreation, navigation and the management of fish stocks. At the initial stage these purposes are generally pursued independently, often by quite separate agencies and sometimes at different levels of administration. As the impacts of developments for individual purposes increase, so pressure for co-ordination

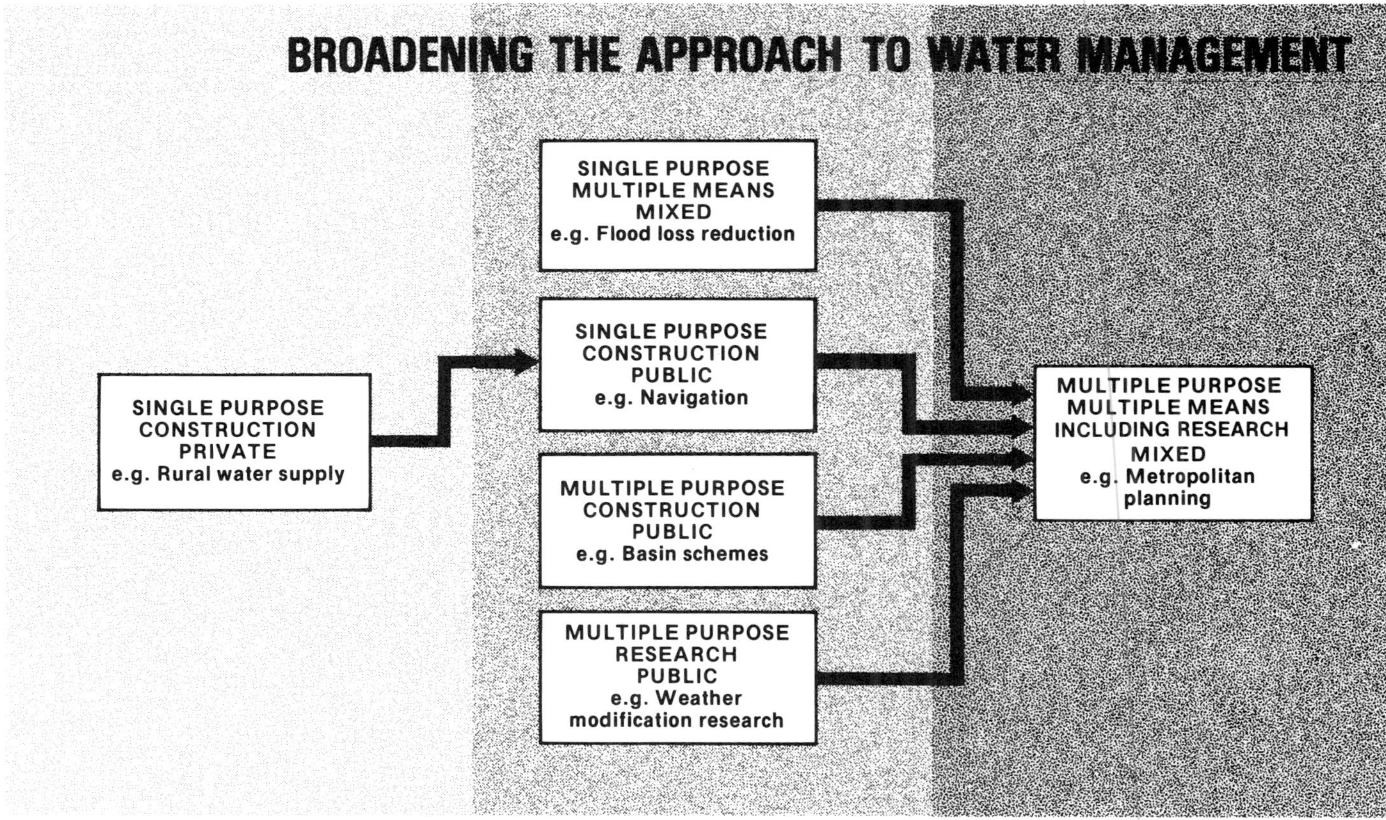
BROADENING THE APPROACH TO WATER MANAGEMENT
SINGLE PURPOSE
CONSTRUCTION
PRIVATE
e.g. Rural water supply
SINGLE PURPOSE
MULTIPLE MEANS
MIXED
e.g. Flood loss reduction
SINGLE PURPOSE
CONSTRUCTION
PUBLIC
e.g. Navigation
MULTIPLE PURPOSE
CONSTRUCTION
PUBLIC
e.g. Basin schemes
MULTIPLE PURPOSE
RESEARCH
PUBLIC
e.g. Weather
modification research
MULTIPLE PURPOSE
MULTIPLE MEANS
INCLUDING RESEARCH
MIXED
e.g. Metropolitan
planning

Figure 1.1

among them grows. Ultimately, total integration of purposes under a single agency may be seen as the most appropriate solution. A Ministry of Development of Water Resources or a Water Resources Commission might be formed to this end, depending on the situation and on the functions to be undertaken. There are examples of such bodies at central levels of government (as in Hungary and Israel) and at state or provincial levels (as in the United States, Canada and Australia). Beyond that there may be bodies established to achieve integration of purposes at the regional or local level. The Regional Water Authorities in England and the Tennessee Valley Authority and the River Basin Commissions in the United States are multiple-purpose agencies of this kind for regional water management.

Generally, there is an evolutionary process of integration of purposes extending over a long period of time. Typically it begins with the linking of water supply to waste disposal and then proceeds to encompass the control of water pollution. Other purposes, such as flood control, the management of fisheries or the provision of water-based recreation often enter at a much later stage. Sometimes integration may begin with a consolidation of one of the latter functions with one of the former. Management of fisheries and the control of water pollution provide an example.

Besides integration of purposes, it may also be desirable to consolidate the various functions of water management into a single agency. Thus, responsibilities for data collection, planning, regulation and development might be better concentrated in one body than shared among several. Experience in various parts of the world suggests that there is an evolutionary process in which the number of functions multiplies over time and in which the magnitude and sophistication of each of them expands (Craine, 1969; United Nations, 1974). At an initial stage of development, for example, the focus may be on a single purpose, such as water supply, and the planning unit may be a village or a small town. The collection of data may be confined to information on streamflow in the local area and planning may deal with little more than the search for appropriate points of withdrawal. At higher levels of development an increasing range of types of data may be needed, not only on physical aspects but also on environmental and social dimensions.

This evolutionary process is described in Figure 1.2 which shows that, as development becomes more sophisticated, the number and magnitude of functions increases and there is a progressive integration of the various activities. The diagram is intended only to illustrate the process. It does not include, for example, all the possible functions. Moreover, the order in which they appear at a given stage of development may differ in an actual situation from that shown in Figure 1.2. In some instances, for example, the process may begin with the collection of data and then move directly to regulation without a planning function being included.

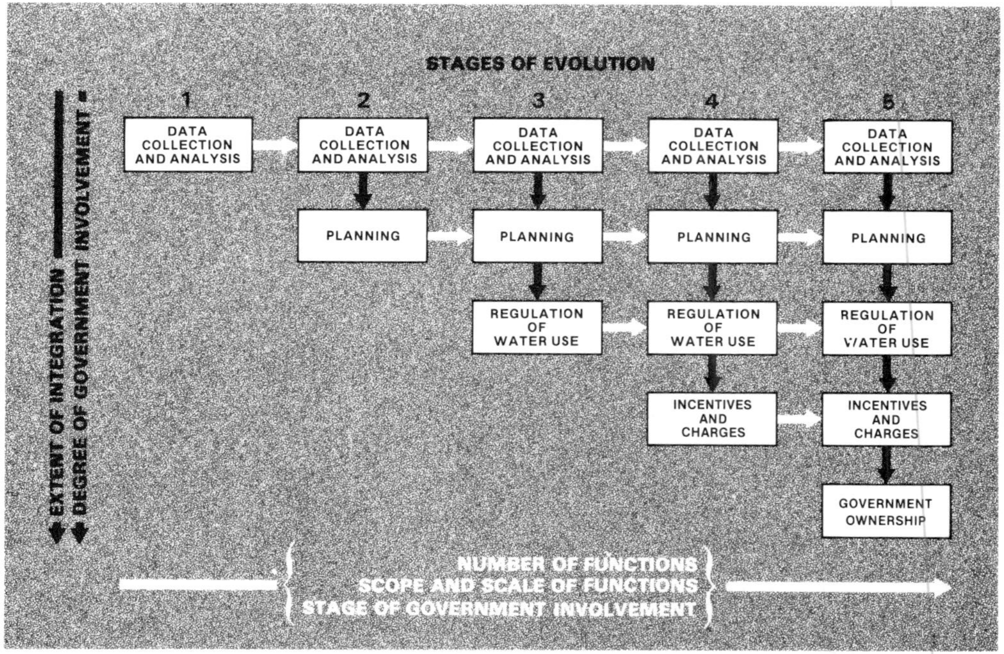

Figure 1.2 The evolution of functional integration.

One important implication of the process described above is that higher stages of development tend to be accompanied by an expanding degree of involvement by government. This occurs as increasing reliance is placed upon public authorities to gather data, sponsor research, undertake planning, allocate resources among competing uses and provide water-related public goods, such as flood control or water-based recreation. It should be noted, however, that although the pursuit of certain purposes of water management, such as the generation of hydro-electricity or flood control, tends to be dominantly in the hands of public bodies, others are typically left to private management, although this is sometimes influenced to an important degree by public policy.

Expanding the Range of Choice

There are usually several ways in which a given water problem may be solved. Generally, however, the range of strategies that is canvassed is much narrower than the number of options that is theoretically available (White, 1961). Thus, problems of increasing demand for water might be dealt with by accepting the possibility of a growing scarcity of water, increasing the supply or improving the efficiency with which existing supplies are used. Table 1.1 identifies some of the specific approaches and techniques that might be considered in this connection (Johnson, 1971). There are also many possible approaches to the problem of deteriorating water quality. These might include the acceptance of deterioration; the imposition of standards and regulations, enforced by the threat of fines; the provision of financial incentives, such as grants, subsidies or tax write-offs; the use of charges for the discharge of effluents or the construction of facilities for treatment (Davis, 1971). Again, problems of flooding might be handled by building control works, treating watersheds, undertaking emergency actions, altering structures to withstand floods, introducing insurance, providing funds for relief and rehabilitation or altering land use (White, 1964).

A review of water management experience in Europe, North America and elsewhere suggests that there is an emphasis on options which involve the construction of additional facilities, while simpler and less expensive alternatives are given only minor consideration. In water supply, for example, the preference has been typically for the construction of new reservoirs and pipelines, rather than for such options as renovating waste water, recycling, or altering charges to encourage more efficient use. In the management of water quality, the emphasis is dominantly on the setting of standards and imposition of fines rather than on the provision of incentives to polluters to reduce discharges of harmful effluents.

Table 1.1 Alternatives for meeting increasing demands

Goals	Approaches	Techniques
Bear shortage or pollution		
Use available supply	Storage	Reservoirs Storm catchment
	Transport	Aqueducts Ground pumpage Motive transport
Increase overall supply	Precipitation inducement	Cloud seeding
	Increase and capture snow- and ice-melt	
Improve water quality	Treat influent	Fresh water purification Desalination
	Treat effluent	Advanced waste treatment
Change water use	Reduce use or alter demands	Price curbs-metering Use restrictions Recycling
	Curb waste	Evapotranspiration reduction
		Seepage reduction
	Alter distribution	Dual supply lines Directed pipelines Bottled water

SOURCE: James F. Johnson, *Renovated Waste Water*, University of Chicago, Department of Geography, Research Paper No. 135, Chicago, 1971.

White (1966) and others (Davis, 1971; U.S. National Academy of Sciences - National Research Council, 1966) have drawn attention to the risks of concentrating on a narrow range of choice. They have noted that not only may the strategy selected not be the cheapest alternative, but that it may be less effective in dealing with the problem. The need to find efficient and effective solutions becomes increasingly important as the competition for available capital grows and as the impact of water development becomes more profound and widespread.

There has been growing pressure in Europe, North America and elsewhere in recent years to introduce legislation, policies and administrative structures which encourage the canvassing of a wider range of potential strategies. This has been especially so in the case of losses from flooding, in which more and more attention is being paid to regulating land use, flood proofing, insurance and emergency measures as complements to, or substitutes for, traditional options, such as the construction of control works or disaster relief. Major legislation, policy and administrative changes along these lines have occurred in the United States (White, 1966), Canada (Tate, 1981; Bruce, 1976) and the United Kingdom (Parker and Penning-Rowsell, 1980; Porter, 1978) in the past two decades.

Similarly, there has been an expanding horizon in the search for solutions to growing shortages of water. Slowly, there seems to have emerged an 'intensive' approach in which the focus is upon making better use of existing supplies, in contrast to an 'extensive' one in which the aim is to find additional water, sometimes hundreds of miles away (Sewell, 1966). Thus, there have been attempts to reduce leakages from distribution systems, which in some instances lose as much as 30 per cent of the supply before it reaches its destination, and efforts to reduce demands by metering (Jenkins, 1973; Hanke and Boland, 1975).

In the management of water quality there is developing a shift from a sole emphasis upon standards and punitive legislation to policies which provide incentives to reduce the volume of effluent placed in rivers. These policies include the use of fees for the discharge of effluent, based on the 'polluter pays' principle (OECD, 1977), and the use of subsidies to encourage the construction of treatment works. The underlying notion in such fees is that the polluter should be given the option of paying the fee or reducing the volume of his discharge (Kneese and Bower, 1968). Revenue collected from the fees could be used to construct the facilities needed to improve the quality of the river. The polluter might decide that, rather than pay the fee, he should alter his production processes to reduce the volume of effluent. The government might assist in this connection by offering rapid write-offs for tax purposes. These approaches have already been adopted in several European countries, including West Germany, France and the Netherlands (Bower et al., 1981; Johnson and Brown, 1976)

Water as an Economic Good

Traditionally water has been regarded as a 'free good', especially in countries with humid climates. In some parts of the world, this status is protected by law. In most countries no charge is made for abstracting water from a stream or for using water bodies for the disposal of wastes. This contrasts considerably with the situation relating to other resources, such as oil and forests, where dues are paid to the government for exploitation or use. Even where fees are paid for the abstraction of water they are generally nominal and bear little relationship to the value derived. As a consequence, there is no basis for the allocation of the resource other than that of prior right; those who make claims first have the highest claim to use. Some legal codes have a list of priorities in allocating water, generally protecting domestic uses and the watering of livestock. Beyond that, there is no particular rationale for the rankings given.

A similar kind of philosophy seems to prevail with respect to water services (Hirschleifer et al., 1960). The prices charged for most other items to which consumers allocate their income reflect the cost of supply and the amount which the consumer is willing to pay. Water, however, is treated differently. Not only is its price usually well below what consumers would be willing to pay, but in many instances it does not truly reflect the actual cost of supplying the water. Most water utilities use average costs as a basis for charging. As a result, those who live in isolated districts or who place demands on the system at peak periods are subsidized by those who live in areas that are less costly to supply or who make little or no contribution to peak demands.

Various economists, geographers and others have pointed out the inequalities and the inefficiencies of the present system and have called for the introduction of pricing schemes which endeavour to reflect the costs involved and the values derived, namely an approach based on marginal cost pricing. Their rationale is that this will lead to substantial reductions in waste and in the investment required for water services, and that those who contribute most to costs will bear their fair share of the burden of expenses. Case studies undertaken in the United States (Hanke and Davis, 1974) Canada (Sewell and Rouesche, 1974) and the United Kingdom (Rees, 1974; Herrington 1981) emphasize these claims.

There has also been pressure for a more rigorous appraisal of the benefits and costs of schemes of water development. Not only are such schemes becoming very expensive, but they compete with other, equally important uses of capital and other resources. At one time a financial analysis, involving an appraisal of the likely revenues against the capital, operating and maintenance costs, was regarded as sufficient. Now it is important to review not only these aspects, but also the likely impact of a proposed scheme on the economic and other activities of a region, and the extent to which it would provide the desired services at a cost lower than other options. As a consequence, the use of benefit-cost analysis and other techniques of economic evaluation has become standard practice in water planning in many parts of the world (Sewell, 1973). In some countries, such as the United States and Canada, the evaluation now goes beyond economic considerations to include environmental and social impacts (O'Riodan and Sewell, 1981). At the same time, there has been growing interest in hindsight evaluations, aimed at determining the extent to which given projects or policies of water development have attained their objectives or have had unforeseen consequences. Such evaluations are seen not only as a means of assessment performance but also as a contribution to the planning of future development (United Nations, 1976; Day, 1974).

Public Involvement

In many parts of Europe and North America there have been mounting demands in recent years for a more direct role of the public in water planning and policy-making. Such pressure results from a concern that the individual is becoming increasingly remote in many cases of decision-making and from the fact that planners and policy-makers have sometimes failed either to assess correctly the preferences of the public at large or to predict the consequences of the projects they have selected. The result on occasion has been that the public has rejected schemes which planners believed were in the public interest, or that certain groups had suffered unforeseen losses of income, habitat or amenity.

Springing in large part from the environmental movement of the late 1960s and the early 1970s, large numbers of interest groups were formed in Europe, North America and elsewhere, aiming to influence more directly the formulation of policies for water management (Sewell and Coppock, 1977). The response from governments has varied considerably, but there has been a general recognition that the public had a right not only to be informed but also to voice its views more directly than had traditionally been the case in water management. In some instances, notably in the United States, there have been major efforts to provide opportunities for a more meaningful role for the public in planning and policy-making (Pierce and Doerksen, 1976) involving experiments with numerous techniques, ranging from surveys of public opinion and letters to editors and politicians, to town meetings, workships, seminars and task forces.

Although views differ sharply on how far public participation should go (Wengert, 1976) and which methods are the most effective (Sewell and Phillips, 1979), there is substantial evidence not only that the public desires a wider role, but also that it can make an effective contribution to planning and policy-making (Pierce and Doerksen, 1976). The challenge is to determine how, in given cultural circumstances, such public inputs can best be obtained and put to use (Sewell and O'Riordan, 1976; O'Riordan and O'Riordan, 1979).

Public participation may be more widely interpreted to include formal and informal consultation of various kinds. Parker and Penning-Rowsell (1980) identify three kinds of groups influencing policies and plans for water management in England and Wales: professional bodies and associations of employees; organizations with a direct financial interest in the outcome; and environmental groups. They see such groups exerting influence in three main ways: through membership on advisory or policy-making bodies; through advice to government, whether in response to formal invitations or through informal contact; and through evidence at public inquiries.

ASSOCIATED DEVELOPMENTS

These themes have occurred to varying degrees in different countries. Where the underlying concepts have been fashioned into laws, policies and administrative structures, they have led to one or more of the following associated developments: (a) an increasing involvement of government in planning, regulation and development of water resources; (b) the adoption of the river basin as a basic unit for water management; (c) a broadening of the range of specialists involved in the water industry; and (d) a growing sophistication in water planning.

Increasing Government Involvement

Although there has been considerable variation in the extent to which water management has been in public or private hands, the past two or three decades have witnessed a major expansion of involvement of government in water management. This has resulted in part from the fact that conflicts have increased in number and complexity, and governments have been obliged to arbitrate among them and, in some instances, to institute regulations. Beyond this, they have been called upon to furnish a variety of services which typically are either not provided or inadequately provided by the private sector. These public goods include such services as flood control, water purification plants, navigation facilities, or opportunities for outdoor recreation.

Not only has government involvement increased, but it has generally been at the senior levels that the greatest expansion has occurred. This has been because responsibilities have gradually moved from local levels of administration to regional and national ones in order to obtain the benefits of co-ordination and economies of scale. The experience in England and Wales is illustrative, where there has been a progressive move from a myriad of small, single purpose, local agencies to ten regional, multiple-purpose authorities. A similar though less comprehensive development has occurred in France with the establishment of six river basin agencies.

The River Basin as an Areal Unit for Planning and Management

In many countries the river basin has come to be accepted as the basic areal unit for the management of water resources (Teclaff, 1967). In some instances, such as the United States, Canada, Hungary and a few countries in the Third World, the provision of certain water-related services is organized by river basins in some parts of the country (United Nations, 1976). In a few instances, notably in England and Wales, the entire process of water management is undertaken on the basis of river basins. Whilst such a development has clearly had many advantages, it is still viewed with scepticism by those who believe that administrative units, such as cities or countries, constitute more logical and manageable alternatives.

Water managers in general tend to argue in favour of the watershed or river basin. They see the latter as an organic entity, such that a development or alteration in one part of it ultimately affects all other parts, resulting in what economists have called 'externalities'. These might be unforeseen losses, as when the discharge of effluent by a factory upstream leads to increased costs of treating water in a plant downstream, or windfall gains, as with the storage reservoir upstream which makes possible increased output of power at generating stations downstream. Only by managing the water resource by river basins, proponents argue, can such externalities be internalised (Kneese and Bower, 1968). Moreover, when the development of water resources is being used as an instrument to foster economic and social development, the river basin would seem to be the appropriate unit for management (United Nations, 1970).

There have also been arguments against the use of river basins as management units. Many of the physical interdependencies and technological spillovers, for example, are not as pervasive as is sometimes assumed. The effects of dams and the impacts of waste disposal frequently become dissipated downstream, especially when the river system is a large one (Maass, 1962). In addition, the lack of congruence between the boundaries of surface and groundwater resources, and between river basins and political boundaries has made it difficult not only to manage water resources but also to relate such management to other government functions (Martin, 1963). Beyond this, and especially in large metropolitan areas, the water resources of a given basin may become inadequate to serve all the needs of the area. In more developed regions, the demand and supply patterns relating to water, electricity, agriculture and recreation seem increasingly to cut across the boundaries of river basins, leading to interdependencies between basins; people from one basin use the resources of other basins, either by moving into the latter temporarily or by transferring water into the former. In such circumstances, therefore, the size of the management unit will necessarily expand beyond the boundaries of a single river basin.

Where water development is the major focus of economic activity, the river basin might well be the most appropriate areal unit for its management. With continued expansion in the size of cities, however, and the increasing interdependence between town and country, the city and its environs now form a closely knit socio-economic unit. In addition, problems of water management in urban areas have grown considerably in scope and magnitude. Several observers have suggested that such problems cannot be handled satisfactorily if the river basin is the unit of management and that some other kind of unit is needed (Kelnhofer, 1968; MacPherson, 1970). Some have proposed that, where urban centres cover a particularly large area, the boundaries of the metropolitan region should be used as the boundaries for water management (Kelnhofer, 1968). A few go even further and suggest that water functions should be integrated with other functions, such as redevelopment, transport planning and housing, and managed by a metropolitan authority (Bollens and Schmandt, 1965). There is, however, no complete agreement on this matter, as the wide variety of units now in use clearly testifies.

Broadening the Range of Specialists

Traditionally, water management has been the business of a small group of specialists, principally engineers, chemists, biologists and, to a more limited extent, lawyers. As the perspective has broadened, however, the need for ideas and skills from several other fields has become apparent. Recognition of water as an economic good, for example, has emphasized the importance of including economists in management teams. The impact of water development on the environment and on social and economic relationships has stressed the value of such fields as ecology, geography and sociology. At the same time, research is becoming recognized as an important management tool, bringing an increasing demand for specialists in that area too.

There have been some significant moves in several countries to accommodate these needs. This has been especially so where the comprehensive planning and development of river basins has become an established fact, as in the United States, Canada, and, to a lesser extent, England and Wales. Elsewhere the traditional disciplines continue to dominate.

Increasing the Sophistication of Planning

Until relatively recently, the planning of water resources in most countries was a small-scale activity, usually undertaken on an intermittent basis. Typically, it was performed by a single individual or a small group. Its scope reflected the perspective of the day. In most instances it consisted of little more than examining a few sites which might be developed for reservoirs and designing the necessary control works and pipelines. Future demands were typically forecast on the basis of extrapolations of past experience. The time horizon was usually short, seldom extending beyond ten years.

There has been a dramatic change in attitudes towards planning. Most contemporary agencies responsible for water management now have a planning division. Those who work in them are often specialists and tend to work on planning on a continuous basis. Their responsibilities extend well beyond the assessment of streamflows and dam sites, and they are increasingly involved in the preparation of long-term plans. In many instances they come into contact with officials of other agencies and sometimes with members of the public as well.

As the perspective of planning has broadened, the number of dimensions to be considered has grown considerably. An increasing range of tools and techniques has been developed to assist in this task, including various forms of benefit-cost analysis, engineering economic analysis, environmental impact assessment and social impact assessment (Sewell, 1973). In some instances, use is made of various types of systems analysis.

THE NATURE AND PACE OF CHANGE

From the investigations of institutional change that have been made so far, it is possible to make several broad generalisations. Firstly, institutional change seems to take place slowly and incrementally, often in non-sequential steps, i.e., by what Lindblom and Braybrooke (1963) called 'disjointed incrementalism.' There are seldom wholesale modifications of an entire institutional framework; instead, some parts are altered whilst most of the structure remains the same. Thus, there may be important changes in the legislation dealing with water pollution but little or nothing is done to the laws that concern water supply. Again, although there may be changes in legislation, these may not be accompanied by modification to policies or the structure of the agencies.

Secondly, change tends to be conditioned by two broad sets of factors, internal and external, which may either promote change or retard it. The first set of factors comprises those internal to the system, such as the goals and objectives of the organisation, the role of the various actors and their perceptions of the problems and possible solutions, and past precedents. The external factors are often only marginally related to water management as such, but help to shape its institutions and policies; they include such matters as the legal framework, the state of the economy, social and demographic trends, and major hazards, such as fires, earthquakes and wars. At one time, internal factors may be dominant, at another time, external. In any case, crises are frequently the stimulus to change and the means whereby it is effected.

Thirdly, because of its nature, water management is profoundly conditioned by the perceptions and attitudes of professionals, especially engineers (although this is, of course, only a special case of the previous point). Professionals play a key role in identifying not only problems and issues but also the potential strategies to deal with them. A consequence is that the issues are generally seen as technical rather than human and there is considerable resistance to changes that would take water management, even if only partially, out of the hands of engineers.

Lastly, there is a tendency for some areas to be leaders and others to be followers. When the government offers subsidies for the construction of a plant for the treatment of sewage, for example, some municipalities will respond immediately, whilst others, with comparable problems, will ignore them. A study of the pace of adoption of various policies in English water management showed that there was a fairly rapid adoption of charging schemes, but the pace of expansion of hydrometric schemes was much slower; and whilst some authorities completed their initial periodical surveys on time, many did not (Sewell and Barr, 1978). The reasons for such differences in the rate of institutional innovation in resources management in general and in water resources in particular are not well understood. Major social costs may, however, result from inaction: a better way of doing things may be available but ignorance or overt resistance may result in its neglect.

THE SCOTTISH EXPERIENCE

Scottish experience, particularly in the period since the Second World War, provides an opportunity to examine the pressure for, and response to, institutional innovation in water management. To evaluate that experience it is necessary to appreciate a number of features of the Scottish scene that derive in part from features of its geography and in part from the history of the Scottish nation within the United Kingdom.

Firstly, Scotland, by virtue of its location, terrain and climate, is very well endowed with water resources, in both absolute and relative terms (Scottish Development Department, 1980). Over the country as a whole, precipitation averages about 1400 mm (55 inches) per annum and does not vary greatly seasonally. After allowing for evaporation and other natural losses, there is a total of 71 million megalitres (15,600,000 million gallons) a year, or an average of 200,000 megalitres (45,000 million gallons) a day. Since the population of Scotland is only 5.1 million, or about a tenth of that of Great Britain, each individual has theoretically 39,000 litres per day (8,600 gallons) available to him. Actual consumption per capita is much less, at 450 litres per day (99 gallons), though it is still high and compares with some 10 litres (2.19 gallons) in some parts of Africa. With such an endowment, problems might not be expected. That they do occur is largely due to the concentration of much of that population in the towns and cities of the Central Belt (Figure 1.3). This is the most highly industrialized part of Scotland where are concentrated many of the heaviest water-using and polluting activities, notably coal mining, distilling, iron and steel making, paper making and the manufacture of petro-chemicals. In the sparsely-populated rural areas, problems of water management are relatively few and are local in nature; it is in the Central Belt that most difficulties occur and are aggravated by problems associated with ageing industry, poor housing and above-average levels of unemployment.

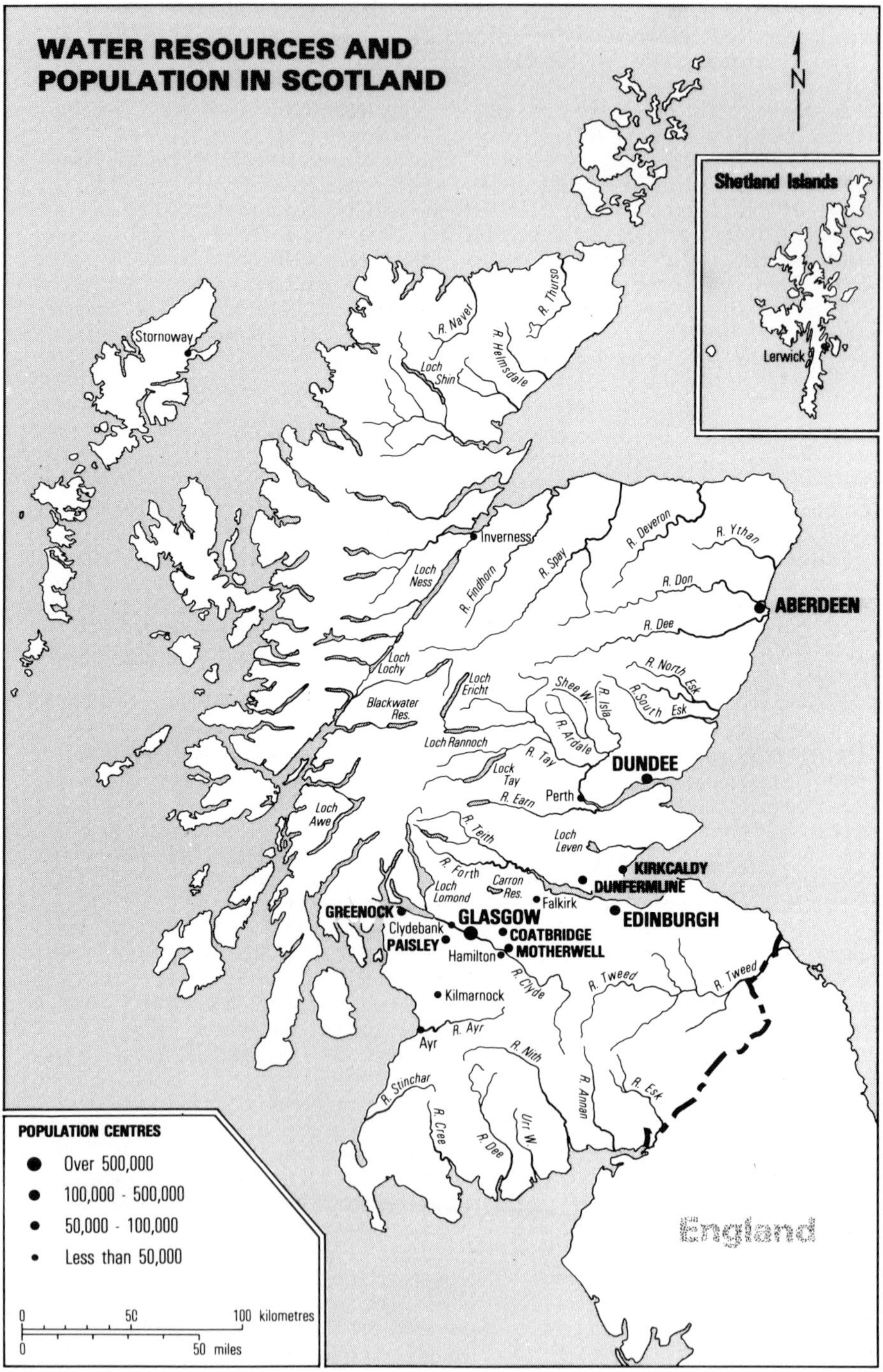
WATER RESOURCES AND POPULATION IN SCOTLAND
N
Shetland Islands
Lerwick
Stornoway
R. Naver
R. Thurso
R. Helmsdale
Loch Shin
Inverness
Loch Ness
R. Findhorn
R. Spey
R. Deveron
R. Ythan
R. Don
ABERDEEN
R. Dee
R. North Esk
R. South Esk
Shee W.
R. Isla
R. Ardale
Loch Lochy
Loch Ericht
Blackwater Res.
Loch Rannoch
R. Tay
DUNDEE
Lock Tay
R. Earn
Perth
Loch Awe
R. Teith
Loch Leven
R. Forth
KIRKCALDY
DUNFERMLINE
Loch Lomond
Carron Res.
Falkirk
EDINBURGH
GREENOCK
GLASGOW
Clydebank
COATBRIDGE
PAISLEY
MOTHERWELL
Hamilton
R. Clyde
R. Tweed
R. Tweed
Kilmarnock
R. Ayr
Ayr
R. Nith
R. Stinchar
R. Annan
R. Esk
R. Cree
R. Dee
Urr W.
England
POPULATION CENTRES
Over 500,000
100,000 - 500,000
50,000 - 100,000
Less than 50,000
0
50
100 kilometres
0
50 miles

Figure 1.3

Although the total population has remained remarkably stable over the past forty years, it is these social and economic problems of the Central Belt that have been mainly responsible for shifts in demands for water services, primarily within the Central Belt. Attempts to solve them have led to the rehousing of substantial numbers of people, both in housing estates on the periphery of existing towns and cities and in New Towns, established mainly to deal with the problems of Glasgow. Similarly, policy (both national and local) has sought to encourage new industries to replace declining traditional industries and these, generally footloose, have sometimes been major consumers of water. Improved housing and improved living standards have led to increased demand for water services; new locations for housing and for new industry have led to a redistribution of that demand.

Two other characteristics merit particular note at this stage. First, Scottish water supply has, unlike that in many a developed country, long been a public service. Private companies did play a more important part in the development of water supplies in the 19th century, but by 1945 all but a handful had disappeared and these soon followed. The provision of sewerage and the control of water quality have, of course, always been public services. It is true that some aspects of Scottish water management remain in private hands, notably fishing and other forms of recreation, land drainage and, to some degree, flood control; but these are all of much less importance in Scotland than elsewhere, notably in England and Wales. Institutional changes in Scottish water management have thus very largely occurred within the public sector.

The second aspect relates to the degree of control exercised by Scottish institutions over Scottish affairs within the United Kingdom. Scotland has a separate legal system and many aspects of government require separate legislation, enacted through the British Parliament in Westminster. There is also a considerable degree of administrative devolution, though this did not become fully effective until the Reorganisation of Offices (Scotland) Act of 1939. Scottish legislation and Scottish institutions have often resembled those adopted in England and Wales, but have been modified to suit Scottish conditions and Scottish problems. They have also tended to lag behind, partly because problems have often appeared less pressing than in England and partly because of difficulty in finding space for Scottish legislation within the Parliamentary timetable. Moreover, the scale of government in Scotland has tended to make relationships both within central government and between it and local government much more intimate than in many larger countries. The lack of direct control over Scottish central institutions (since members of Parliament are in London most of the time and Scottish affairs are only one concern of the British Parliament) has tended to give civil servants, both administrative and professional, more influence over the shaping of policies and institutions than might otherwise be the case.

These considerations provide another context for this book, which traces the way in which the various institutions concerned with water management in Scotland have evolved over the past half century. Chapter 2 describes how the different levels of government became increasingly involved, particularly in water supply, sewerage and the control of water quality. It shows that, whilst the antecedents of the present structure may be traced to the problems of urban and industrial growth in the 19th century, major deficiencies did not begin to appear until the 1930s. It was at that stage that a basic division on both the needs and the philosophies of town and country planning became evident.

Although the need for change was recognized in the 1930s, it was not until some twenty years later that substantial changes in institutions occurred with respect to the management of water quality and another decade before there was a major alteration in the approach to water supply. Chapter 3 describes the various pressures that led to changes in the legislation, policies and administrative structures concerning water supply and details the responses that followed. In a like manner, Chapter 4 outlines the factors that led to changes in the approach to the management of water quality in Scotland, aimed particularly at the problems presented by pollution in the Central Belt. Chapter 5 describes the various arguments that were made for and against integration of purposes and consolidation of such functions when the reorganisation of water management in Scotland was being discussed in the late 1960s. It shows that the stimulus to such changes came primarily from external forces, particularly the reorganisation of local government, rather than from pressures within the industry itself.

Much of the present institutional structure for water management in Scotland was established in 1975, as a consequence of the Local Government (Scotland) Act 1973. Chapter 6 describes the manner in which this legislation has had little impact in terms of expanding the range of choice of strategy in Scottish water management. Various factors constrain any marked deviation from conservatism of method although it is clear that there is some scope for initiative when the right men find themselves in the right place at the right time.

Chapter 7 concludes the volume with an analysis of the institutional innovations that have taken place in Scotland since 1945, describes the roles that various factors have played in the evolution, and comments on the extent to which Scottish experience differs from that elsewhere.

2

Antecedents

Few problems in water management are entirely new and those facing Scottish water managers are no exception. Indeed, the key issues with respect to water supply, sewerage, treatment of sewage and control of pollution were apparent well before the Second World War. Serious problems had arisen because of an inappropriate institutional structure, which had grown up piecemeal, largely in response to rapid and highly-localised urbanisation and industrialisation before the First World War. In the interwar period, problems became increasingly acute and pressures on water supply and on waste disposal services were particularly evident on the urban fringe. Moreover, the context of water management was changing as an emphasis on public health gave way to specific policies with respect to housing and planning.

Water management in Scotland was, from the beginning, essentially local. As in many other parts of Europe, the modern concept of water supply had its origins in the appalling conditions in large towns in the mid-nineteenth century. Indeed, the creation of one of the world's first modern water supply systems, Glasgow's Loch Katrine works, can be attributed directly to major epidemics of typhoid and cholera that affected the city. Such outbreaks stimulated a movement to improve the conditions in which people lived by providing them firstly, with adequate supplies of wholesome water and a satisfactory means of waste disposal, and secondly, with better housing. Central government was at first content to allocate responsibilities to local authorities, confining itself to ensuring that minimum standards were met. The provision of supplies of wholesome water and proper sewerage quickly became a symbol of modern civilised living and, as such, spread from the major urban areas to smaller and more rural centres of population.

This chapter outlines the initial allocation of functions to authorities and provides an account of the problems that emerged from the administrative fragmentation of city, town and village. The changing context of water management is also described briefly, namely, the reform of local government in 1929, the need for extensive public intervention in the field of housing, and the introduction of measures to alleviate chronic unemployment. This review sets the scene for the emergence of specific policies for water management which are considered in succeeding chapters.

THE INITIAL ALLOCATION OF PUBLIC HEALTH FUNCTIONS TO LOCAL AUTHORITIES

The pattern of authority in urban areas that could respond to the need for action in the interests of public health was established in the 1830s in which burghs were empowered to accept a 'police system', to be administered by 'commissioners of police'. The right to adopt this system was extended gradually to most 'populous places', as were the duties of police commissioners, which were outlined in a series of Burgh Police Acts dating from 1850 to 1892 and included the improvement of water supplies and the prevention of infectious diseases (Adams, 1978).

Progress in similar fields in rural Scotland was delayed by the lack of any similar responsible body. Until the enactment of the Local Government Act in 1889, at which time County Councils were formed, parochial boards were responsible for public health services. After 1889 the larger counties were divided into 'district' divisions for the administration of roads and public health (spheres of overlapping interest between parish and county) and services were administered by District Committees comprising the county councillors for the district plus one representative of each parochial board (later parish council) and each burgh in the district (Mackie and Pryde, 1935). In 1929 parish councils were abolished and the district committees of county councils were replaced by District Councils, though the full County Council assumed responsibility for all social services, education, public health, roads, housing, slum clearance and town planning. Although the rights of the small and large burghs to operate water undertakings continued undisturbed, local government reform in 1929 meant a loss of function for these councils in other directions. Within this changing framework, systems of water supply developed at different speeds in different places.

VARIATIONS IN THE ABILITY TO PROVIDE A WATER SUPPLY

The extent to which supplies had been provided in adequate quantities and satisfactory quality seems to have been related to the size of the responsible authority. The cities clearly led the field and by the 1870s, most of the larger towns were adequately if modestly provided with water by means of small projects that required frequent augmentation as supplies were extended to more houses and as the urban population grew (River Pollution Commissioners, 1872). Whilst Edinburgh and Glasgow were working at distances of from fifteen to thirty five miles (24 to 56 km) to use large resources which would meet long-term requirements, the other towns of central Scotland seem to have sought their supply no further than 5 miles (8 kms.) distant, at least at first. Whilst most supplies were provided by public authorities, in some areas, such as Perth, Airdrie and Coatbridge, there were private companies; these had also been responsible for the initial but unsatisfactory supplies of Edinburgh and Glasgow. These private systems were, however, progressively taken over by municipal authorities, partly because they could not raise the necessary capital for investment, but also because the spirit of the age favoured the provision of public services by public monopolies.

A third group of smaller urban communities, such as Forfar, Linlithgow and Musselburgh, relied on a series of local wells and springs. Musselburgh had taken its water from the River Esk but, by the 1860s, this had become impracticable because of industrial development upstream (principally paper manufacture) and the town had had to resort to shallow wells in a thin bed of gravel on top of clay. Dirty water from streets and sewers soaked to this bed, so that the supply could be described as dangerous and the high incidence of diarrhoea in the town attracted adverse comment (River Pollution Commissioners, 1872).

If conditions were bad in the small towns, they were worse in the mining and industrial villages scattered throughout otherwise rural areas. Furthermore, because an institutional structure to tackle the problems was lacking until the end of the century, they remained unsatisfactory for longer.

Under the Public Health (Scotland) Act of 1867, the basic administrative unit for the provision of supplies outside the towns was the Special Water Supply District (SWD). The principle underlying the formation of such districts was that the services should be financed locally by means of a special rate. This approach, however, brought several problems. For example, a SWD was formed in 1882 for the industrial village of Chapelhall, in the Clyde basin not far from the city of Glasgow, but the value of rateable property within its boundaries was too small for it to undertake projects on its own account. The growing population, therefore, continued to carry water from mineral bores and a channel bed (Lambie, 1919).

Until the Public Health (Scotland) Amendment Act of 1891, an upper limit of 30d (£0.125) per £1 was imposed on the rates levied in SWDs, and this was often inadequate to provide a water supply. It is true that the Act did allow County Councils to subsidise SWDs through the imposition of a public water rate, but this was not to exceed 3d (£0.0125) per £1. Few councils took advantage of this provision (DHS, Committee on Scottish Health Services, 1936a).

Although the Middle District of Lanarkshire was unusual in using its powers under that Act to work towards providing a proper supply for the whole of its area, it does illustrate some of the problems which water authorities faced in these early years.

In 1892 the District had promoted legislation to authorise a merger of the nine existing SWDs in its area and the construction of two reservoirs to assure their joint supply. Consultants had identified a suitable site at Glengavel, but the contractors could not reach bed rock and other consultants recommended that the dam should be resited upstream. To do so required a second Act and a third was needed to authorise the building of a railway to the construction site and to clarify the District's powers to levy rates. Finally, owing to various delays, a fourth Act was required in 1902.

In 1907, the neighbouring district committee took over a private company, the Busby Water Company, and it was agreed that the Company's territory in Lanarkshire should become the responsibility of the Middle District. The latter's area was simultaneously contracting as the boundaries of the growing towns in Lanarkshire were extended, though the County Council resisted such moves, sometimes successfully. By 1910, demand had outstripped supply. One option was to proceed with the original intention of construction reservoirs at Glengavel, but another was to investigate the cheaper alternative of taking bulk supplies from one of the burghs. A review by consultants showed, however, that Motherwell was the only possibility and then only for a limited period. Their advice was to retain the existing administrative structure, but to develop a major scheme based upon a reservoir in the headwaters of the Clyde.

An Act authorising this scheme was passed in 1913, though the scheme was not, for a variety of reasons, completed until 1919. Among these was a decision to extend the District's statutory area to include two parishes in Dunbartonshire where existing sources were quite inadequate (with inhabitants in some parts buying water from itinerant hawkers during a drought). The Middle District had agreed to supply them after water had become available following the extension of Motherwell's statutory area for water supply to the whole of the burgh in 1914. In some ways this can be regarded as a precursor of the concept of 'added areas' that was to be adopted in the 1970s, following the passage of the Local Government (Scotland) Act, 1973.

Although the Middle District was atypical in amalgamating the SWDs and in using the discretionary powers of the 1891 Act, its experience in other respects appears to have been quite common with regard to three matters: firstly, in frequently changing its boundaries of supply, even to the extent of taking over responsibility for parts of the areas of other authorities; secondly, in the frequency with which recourse had to be made to private legislation, the consequent jumble of enactments relating to the water undertaking and the time taken by legal processes; and lastly, in the fact that as early as 1910 it could be said that the opportunity to exploit water resources in a more rational manner had already been lost because of the division of responsibility between town councils and rural authorities.

The result was an astonishing array of different agencies, involved in the supply of a comparatively small area. In addition to the Middle District and the three burghs within it, supplies were also provided by Glasgow Corporation, the Airdrie and Coatbridge Water Company and the Busby Water Company. When the last was taken over by Renfrew County Council, new arrangements were made to supply its territory in Lanarkshire, but when the Airdrie and Coatbridge Company was taken into public ownership, to form the Airdrie, Coatbridge and District Water Trust, no such rearrangement took place. The Middle District came to the aid of Dunbartonshire in supplying two of its outlying parishes, whilst parts of its own territory (and rateable assessment) were lost through the extension of burgh areas to accommodate suburban growth. By 1910, five authorities (the Middle District, three burghs and Airdrie District Trust) were involved in the supply of this relatively small but important part of West Central Scotland, though this should be compared with the fourteen of twenty years before (9 SWDs, two water companies and three burghs).

Elsewhere, no attempt was made to substitute something better than the system of Special Districts for the supply of water to rural parishes. A Committee on Scottish Health Services reporting in 1936 therefore found it necessary to consider the system in some detail because of the problems it raised. There were 1,700 special districts in Scotland at that time, each with separate rates, and any one locality might be served by several, each for different purposes, including not only water supply but also sewerage, street cleaning, lighting and public parks. It was difficult to understand why housing was provided by the county as a whole, while water and sewerage, essential to its provision, might be, and usually were, provided only by way of special districts. On the other hand, according to evidence received by the Committee, most of the distinctly rural counties were opposed to the abolition of such districts. It was only in counties that were largely industrial, such as Lanarkshire, and in which a service had been provided over a large part of the county that it was comparatively easy to abolish all special districts (DHS, Committee on Scottish Health Services, 1936a).

SEWERAGE AND SEWAGE TREATMENT

A similar array of problems affected the disposal of dirty water. The Committee on Scottish Health Services found that, whilst the cities and larger towns had reasonably adequate sewerage, in many other parts of the country the position was unsatisfactory. A memorandum submitted by the Sanitary Inspectors Association summarised the situation in the early 1930s:

> Progress in providing proper drainage facilities for the towns and villages in Scotland had not been so marked as is the case for water supplies. Many rural areas are without drainage due to the heavy costs involved in providing it. In such areas sanitary progress is consequently at a standstill. In a number of towns and villages the sewers are inadequate to deal with the volume of sewage they are required to carry, and the sewage works are of antiquated design and incapable of dealing efficiently with the sewage. Here again the cost is the obstacle which prevents improvements being carried out.
> (DHS, 1936A, pp. 122-123).

Clearly the difficulty in many areas was to get waste water into pipes, no matter what happened to it after that. The main problem in disposing of sewage was its cost, for the construction of efficient purification or other works for the disposal of sewage and their maintenance exceeded the cost of the rest of the service.

As with water supply, there was a multitude of small schemes in operation where larger schemes covering wider areas would have been more economical, more efficient and more easily maintained to a high standard. The Royal Sanitary Association had again summarised the position:

> It is to be regretted that greater advantage has not been taken of opportunities to combine for drainage purposes, particularly on the part of authorities along the course of rivers and large streams. The adoption of separate schemes by each authority in areas suitably placed for the introduction of trunk sewers to the sea or to centralised purification works has resulted in capital expenditure being incurred which would have gone a long way in meeting the cost of regional schemes .... Where individual schemes have been carried out, it is difficult to depart therefrom and embark on large regional schemes on account of the capital expenditures already incurred.
> (DHS, 1936a, p. 125)

The principal obstacle was financial: districts were not extensive, but the capital costs of schemes for both supply and drainage were considerable. The difficulty had been accentuated by the derating provision of the Local Government (Scotland) Act 1929, particularly that under which agricultural land was rated at only one-eighth of its gross annual value. While rate support grants had been introduced to compensate for this loss and to help new development, there was no provision in the Act to secure that any part of the additional money should be used to compensate for the low rateable value of special water districts. Hence in many of them it was simply not financially feasible to extend their supplies in line with growth, let alone to anticipate developments.

LOCAL GOVERNMENT REFORM

The reform of local government in Scotland in 1929 had been, at least in part, engendered by the economic policy of the government of the day. As a fillip to a flagging economy and worsening unemployment, agricultural land and industrial premises had been relieved of all or part of their obligation to pay rates, but this change undermined the financial basis of many rural parishes. The parish as a unit of local government also disappeared and its functions were transferred to new District Councils, whilst the functions of the former district committees (such as the Middle District of Lanarkshire) were transferred to the full county councils. Most of the functions of the small burghs were also transferred to county councils. Many urban communities appear to have reacted strongly to this loss of local control, particularly over education, notwithstanding the fact that they sent representatives to the appropriate committees of the county council. Henceforth, rivalry between the small burghs and the counties became an almost continuous feature of local politics. Indeed, Mackie and Pryde (1935), writing only five years after reform, commented:

> ... the small burghs still claim that the (1929) Act was a piece of unwarranted over-centralisation which has led to increased costs and decreased efficiency, especially in road administration, public health and housing. (p. 30).

The Committee on Scottish Health Services also recognised the impact of the reforms of 1929 as a serious problem, pointing out that, "the transfer of powers from small burghs has created fears in the minds of representatives of large burghs and has fostered jealousy between county and burgh." (DHS, 1936a, p. 294)

Thus, the results of the different development of water services in town and country, itself stemming from the different forms of organisation, were perpetuated in an institutional structure which contained the seeds of conflict between urban and rural authorities. The long tradition of leaving water management functions as much as possible in the hands of local bodies was coming increasingly into conflict with a move towards efficiency through consolidation, in which the urban areas consistently assumed the most powerful role in decision-making.

CONTROL OF RIVER POLLUTION

The evolution of the control of pollution in Scottish rivers requires some understanding of the position of salmon fisheries in Scotland, where the right to fish for salmon in fresh water is legally private property. As a result, the influence of the fishery interests (i.e., salmon - see glossary) has been characterised by considerations of conservation and the preservation of existing rights. Over the years, proprietors have sought and gained significant concessions over compensation water whenever any dam has been made for water supply or the generation of hydro-electricity. The desire to preserve, or if possible improve, existing properties has led the fishery (and angling) interests to act as a principal pressure group promoting the improvement of water quality as well as opposing some schemes of water supply. Indeed, opposition to development from fishery interests ranks with lack of finance as one of the major constraints on the development of Scottish water resources.

The legislative basis of the conservation of salmon has its roots in the Salmon Fisheries (Scotland) Acts of 1862 and 1868. It had been recognised in the 19th century that revolutionary changes in both agriculture and industry were making it increasingly difficult to ensure the survival of salmon stocks, except in the more remote rural areas, notably the North-West Highlands. There was little that the owners of fisheries could do to arrest these changes or to alleviate their effects and landowners were themselves often promoting agricultural improvement and industrial development, especially in the Clyde Basin. Problems arose not only from pollution but also from the despoliation of spawning beds and the building of obstructions to the passage of migratory fish, such as mill weirs. Some protection was afforded by these Acts, though they came too late for many rivers. Nevertheless, they were one reason for the survival of valuable salmon fisheries as owners safeguarded their property by enforcing (largely at their own expense) the fishing codes they contained, including powers over harmful discharges and on obstructions to the free passage of fish. More generally, landowners succeeded in securing the passage of the Rivers (Prevention of Pollution) Act of 1876, the first legislation aimed specifically at the control of pollution, although it proved ineffective in practice. There are three main reasons why this was so. Firstly, the absolute prohibition on new discharges of sewage and trade effluent proved unenforceable; secondly, existing discharges remained uncontrolled; and thirdly, an informal ruling on the part of central government that local authorities could not be prosecuted under its provisions seemed to undermine the entire aim of the legislation.

THE SCOTTISH ADVISORY COMMITTEE ON RIVER POLLUTION PREVENTION (ACRPP)

Over preceding decades several suggestions for changes in the law on pollution had been made, not least by the Royal Commission on Sewage Disposal in 1915. The Commission did not confine itself to technical problems of measuring pollution or of treating effluent. It advanced the view that the legal offence of pollution with respect to sewage should become the act of discharging a liquid that fell below a prescribed standard, which should be defined by statute or by central government at least once every ten years. Yet, despite an almost continuous undercurrent of dissatisfaction over the pollution of rivers, there had been little action.

A significant development in 1928 was the appointment, as a direct response to representations from the British Waterworks Association (an association of water supply undertakings and authorities), of a Scottish Advisory Committee on Rivers Pollution Prevention (ACRPP). Its mandate was:

> ... to consider, and from time to time report to the Scottish Board of Health on the position with regard to the pollution of rivers and streams in Scotland, and any legislative, administrative or other measures that appear ... desirable for reducing such pollution. (DHS, 1931, p.3)

The extent of pollution in Scottish rivers had been established some five years earlier in response to a questionnaire sent in 1922/23 to local authorities and District Salmon Fishery Boards by the Scottish Board of Health (to become in 1929 the Department of Health for Scotland). These findings are shown in Table 2.1.

Table 2.1 Types, Causes and Instances of Pollution Reported in 1922-23

	No Treatment	Unsatisfactory Treatment	Insufficient Treatment	Defect Not Stated	Total	%
Domestic Sewage	360	101	39	38	538	61
Trade Effs	179	43	55	65	342	39
Totals	539	144	94	103	880	100
%	61	16	11	12		

Source: Matthew-Fyfe, "Pollution in Rivers in Scotland", Transactions of the Royal Sanitary Association of Scotland, 1950, p. 34

Domestic sewage was clearly the chief cause of reported incidents. The three categories of treatment may be interpreted as: situations where no works had been provided; situations where works had been provided but these did not perform satisfactorily; and situations where works could not work satisfactorily because of inadequate design or overloading. Whilst the latter two remain today as routine concerns of officers responsible for the control of pollution, the first dominated in the 1920s and 1930s. A major task of local authorities in the 1930s was the provision for the first time of facilities for the treatment of sewage.

The Scottish Advisory Committee used the overview gathered in 1923 to select typical rivers for the study of the kinds of pollution present and its first report, on the Tweed, was published in 1931 (DHS, 1931). The survey of 1922-23 had shown the principal sources of pollution to be domestic sewage and woollen mills, accounting respectively for 54 and 23 of the 88 cases reported. Arrangements for waste disposal made in all of the 16 Special Drainage Districts in the rural parts of the basin had failed to prevent pollution, and in seven of these, there was no provision for treatment of any kind. Even six of the nine burghs in the basin made no arrangements and those made by the other three were inadequate. Moreover, although the county councils (and the burghs until the 1929 reforms) had had a duty to enforce the 1876 Rivers (Prevention of Pollution) Act, these local authorities were themselves serious polluters. The Committee made the general recommendation that local authorities should install and maintain in good working order plant to remedy existing pollutions and suggested that such plant might reasonably be included in the local authorities schemes of work for the unemployed. The trade wastes of the woollen mills could be treated and evidence put before ACRPP suggested that this would be most practicable in combination with the treatment of domestic sewage. The Committee felt that other authorities should follow the lead of the Burgh of Galashiels where the problem of trade wastes pre-dated the construction of a municipal scheme of sewage purification in 1908. From the first, it was provided that the disposal of trade wastes should be done through the municipal system, subject to the manufacturers paying a proportion of the capital and maintenance costs equal to the proportion of trade wastes running through the system. Similar practices have been adopted elsewhere in the United Kingdom.

The treatment of domestic sewage and woollen wastes arising in the Burgh of Selkirk was particularly unsatisfactory. The Burgh was situated entirely on the right bank of the Ettrick Water, but the left bank opposite the Burgh was within the jurisdiction of Selkirk County Council, which could have taken proceedings in respect of pollution at that point. Representatives of the county council admitted that they were fully aware of the pollution coming from the Burgh but had never taken action against offenders, "because of the friendly relationship existing between the County Council and the Town Council" (DHS, 1931, p. 27). The vice-chairman added that he strongly deprecated one local authority taking proceedings against a neighbouring authority or against offenders within its jurisdiction.

Perhaps, because of this attitude, there had been a general failure among the local authorities to discharge their responsibilities under the Rivers (Prevention of Pollution) Act. To a large extent, this was because the authorities concerned have also been the public health authority responsible for sewerage and were accustomed to perceiving pollution only in terms of it being a threat to public health.

HOUSING AND HEALTH

Problems of sewerage, pollution and the adequacy of domestic water supplies were only one thrust in the public health movement. The homes of the Scottish people were also the focus of growing and continuous concern, particularly in respect of overcrowding. The census of 1931 had revealed that 15 per cent of the population lived at densities of more than three to a room, with a further 35 per cent at two to a room (Commissioner for Special Areas, 1939). The position had certainly improved since 1917 when the Commission on the Housing of the Industrial Population in Scotland had identified the tenement as a major impediment to the health and vitality of the urban population. The construction of new styles of housing for renting to the working classes had been recommended to local authorities (Slaves, 1975, pp.245-247). These recommendations formed the bases of the Housing and Town Planning Act, 1919. Between then and 1941, over 300,000 new houses were built in Scotland, 70 per cent of them by local authorities, although it was estimated in 1938 that a further 300,000 were still required. Such development required appropriate servicing, making for virtually continuous pressure on existing infrastructure. In this respect, the provision of adequate water supplies and means of waste disposal was critical.

UNEMPLOYMENT AND INDUSTRIAL DEVELOPMENT

Housing and health were not the only wider issues that had a bearing on water management in the 1930s. The depths of economic recession brought some very high levels of unemployment, e.g., 60 per cent in Wishaw and 54 per cent in Clydebank. Areas where unemployment stood at 40 per cent or more were allocated special assistance under the Special Areas (Development and Improvement) Act 1934. In Scotland, Clydeside and North Lanarkshire (excluding the City of Glasgow) were designated as a Special Area and a Commissioner was appointed to exercise powers of assistance. These were limited: the Commissioner was not allowed to assist private industry directly or to duplicate other government schemes of financial aid, so that the major areas of public works, housing and roads were excluded (McCrone, 1969). In his first year the Scottish Special Area Commissioner spent 90 per cent of his budget on sewerage and his last pre-war report recorded assistance towards 18 schemes of water supply (though from 1937 his main contribution lay in attempting to alter the location and structure of industry by the promotion of industrial estates) (CMS, 5905, 1939).

These water schemes were largely concerned with urgent improvements on grounds of public health but the Commissioner reported that evidence had "come to hand ... connected with water supply which cannot be overlooked" ... "an adequate reserve of water in a particular place at a particular time may be the determining factor in negotiations for attracting new industrial developments." (CMD 5905, pp. 19-20). Further, it has been suggested that the 1930s saw a massive leap forward in the treatment of sewage in areas beyond the major cities and that this was as much due to the provision of unemployment grants as any increase in consciousness about pollution on the part of elected representatives and municipal engineers. Although each of the assisted schemes had to pass the test of urgency on grounds of public health, the result was frequently that the attractiveness of an area to industrialists was enhanced.

Whilst the Commissioner was instrumental in achieving substantial progress in freeing the Upper Clyde estuary from the risk of pollution by sewage and industrial waste, local initiative was not totally absent. Indeed, the first significant move towards cleaning up the Clyde was taken by Glasgow Corporation in 1894. Two years later, a conference of the local authorities upstream of the city was convened by Lanark County Council to consider further steps. Since then 78 purification schemes, costing over £6 million, had been undertaken and 25 of these had qualified for grants. In fact by the 1930s, the Commissioner claimed in his 1938-39 report: "The problem would probably have been completely solved by this time if trade conditions of Clydeside since the war had been better, and if the difficulty of securing sites had not been so acute" (Commissioner for Special Areas, 1939).

As a result of his close involvement in such projects, the Commissioner had several comments and recommendations to make in his last pre-war report. Three facts were of particular importance: the large extent of discharge of raw sewage and untreated industrial wastes; the primitive systems of disposal operating in some special districts; and an almost complete absence of any attempt to foster concerted action by local authorities. On the last point he felt that the time had come for the government to take some strong and immediate steps to deal with the worsening situation.

GOVERNMENT INTERVENTION

All the evidence thus seemed to suggest that government intervention was required not only in the sphere of sewage disposal but also in the prevention of river pollution and in water supply. Yet no suitable agency of government existed. Although a central authority, the Scottish Board of Health, had various responsibilities and duties under the Burgh Police and Public Health Acts, these were not executive but concerned with appeals to the Secretary of State. In 1929, the Board of Health became the Department of Health for Scotland, but it was not a civil service department in the modern sense. The possibility of executive action had to await the assumption of political control over activities and this did not come until ten years later. The publication of the Gilmour Report in 1937 represents a turning point in the history of the Scottish Office and perhaps also of Scotland (Kellas, 1968). It pointed to the impracticability of separating the (political) office of the Secretary of State (in London) from the other Scottish Departments (in Edinburgh). In future, the report suggested, they should all be in Edinburgh and all administrative functions exercised by Scottish agencies should be directly vested in the Secretary of State for Scotland, thus eliminating the quasi-independence of departments. The Reorganization of Offices (Scotland) Act 1939 implemented the Gilmour Committee's recommendations and the new Scottish Office was established in Edinburgh immediately before the outbreak of the Second World War. This measure, together with an increasing professionalism of the departments throughout the 1930s, set the stage for radical change.

It was clear by 1939, that a broader approach to water management was required in Scotland. Programmes of rehousing reaffirmed the need for close links between the provision of water services and the welfare of the community at large. Nor could the needs of industry be ignored. The efforts of the Special Commissioner had revealed the mismatch between a structure of water supply agencies which could, in places, barely afford any new development, and an increasing number of branch plants of foot-loose industires which might be attracted to an area by the immediate availability of supply of water. The contribution of sewerage and water quality was much less explicit, though both ACRPP and the Committee on Scottish Health Services had drawn attention to the problems.

Although a widening perspective was needed, the impetus for political action was missing. At a local level, shortages of water were not yet sufficiently acute, there were few epidemics of water-borne diseases and other issues, such as unemployment, seemed much more urgent. In any case, the worst problems arose from the structure of the water industries and had been aggravated by the reform of local government. Central government, for its part, still regarded water management as an essentially local affair, for which it had a general supervisory responsibility. In the main, it felt it should intervene only where technical advice or coordination was necessary. In sum, the focus remained upon water as a commodity or a service rather than as a key element in economic and social development. Integration of purposes was still a remote prospect.

The recognition of a problem, with which this chapter began, does not imply a trigger for action. With respect to water supply, this came with the drought of 1933, although actions did not generally occur until after the Second World War. With respect to sewerage and the control of pollution there was no such direct stimulus, though in the case of river pollution pressure on water resources in England stimulated debate there and proposals were brought forward for an improved system of water quality control, which would also extend to Scottish developments. Further changes in the context of water management are the subject of the next chapter.

3

Reshaping the management of water supplies

For forty years, from 1933 to 1973, matters of water supply held the centre of the stage in Scottish water management, although it was not until the 1960s that it became an issue of any political significance. The need to reorganise the administration of water supplies was recognised in 1933 when drought highlighted the emerging problems; yet, although attempts were subsequently made to tackle particular difficulties and reorganisation became an object of policy in 1946, reliance continued to be placed on voluntary amalgamation. The basis for reorganisation existed from the late 1940s in the technical appraisals of Scottish water resources made by officials of the Department of Health for Scotland (DHS). Although these findings made a useful contribution to subsequent action, it was not until a change in political priorities in favour of regional development that the political will to action emerged and the possibility that shortages of water might frustrate development was recognised. The result was the Water (Scotland) Act 1967, and it is with that Act and its antecedents that this chapter is primarily concerned.

THE DROUGHT OF 1933

The drought of 1933 clearly demonstrated that in parts of Scotland the highly-localised organisation of water supplies had not kept pace with an ever-increasing demand for water. Reports in 1934 from local authorities on the adequacy of their supplies during the previous year revealed to DHS that 300 of the 878 supply systems reported had proved inadequate (DHS, 1934a). The Committee on Scottish Health Services which was then sitting appointed a special sub-committee to investigate the matter further. It found that some of the areas worst affected had never had a satisfactory supply, whilst others, even in some large urban undertakings, were working on very small margins of safety (DHS, 1934b).

The deficiencies, however, were not due to any lack of water resources. Rather, they were a consequence of the administrative and legal system under which these resources had been developed independently of each other and of their environs, an approach that had led to a good deal of wasteful duplication. The sub-committee argued that the cost of providing water supplies could be significantly reduced if local developments that made full use of available catchments.

All the professional associations consulted by the sub-committee agreed that new arrangements were necessary, whereby the water resources of Scotland could be viewed as a whole and their allocation arranged according to need. If adequate water supplies were to be secured at minimum cost, local government boundaries would have to be transcended, at least for planning purposes. Hence, a complete review of the provision of water supplies would involve not only precise knowledge of existing and potential sources but also consideration of many administrative and legal questions. Gathering facts about sources of supply, the areas that might be served and their needs was a matter for investigation by engineers and this might take some time. A solution for the administrative and legal questions, statutory areas of supply, administering authorities, water rating, water rights and so on, the sub-committee suggested, would depend not only on the technical facts, but also on the willingness of the many interests involved to discuss the matter and to seek an effective solution. The clarification of such questions would inevitably take even longer.

The sub-committee was aware both of the urgent need for action to furnish an improved water supply in several areas and of various institutional impediments to providing it. As a first step it recommended that a technical survey of water resources and needs be put in hand at once. DHS was also aware of the need for a fresh approach. In 1935, it had declared that no local authority should have to provide a new supply when another had ample water which, with a minimum of technical difficulty, could be made available and that no neighbouring authorities should provide independent supplies without investigating the possibilities of co-operation in a joint supply (DHS, 1935). The policy statement was admirable, but DHS had no powers to enforce it.

EARLY LEGISLATION AND THE ROLE OF DHS

In the context of the crisis created by the drought, two Acts of Parliament were passed in 1934, the Water Supplies (Exceptional Shortage Orders) Act and the Rural Water Supplies Act. Both were important precedents. The former empowered DHS to authorise the taking of supplies by Order, if only on a temporary and emergency basis. Hitherto the legal right to take water had always been determined by Parliament and, in the event of individual disputes, ad hoc Parliamentary Committees had determined how water resources should be allocated. Now this right might pass to professional administrators with a sound perspective on the issues involved. The Rural Water Supplies Act empowered DHS to make grants towards the expense of providing or improving supplies in rural areas, subject to its approval of engineering details of each scheme. Most of the projects subsequently approved were local but a few regional schemes were begun, in Dunbartonshire, Dumfriesshire, Easter Ross, Kirkcudbrightshire and Perthshire (DHS, 1935).

The final report of the Committee on Scottish Health Services in 1936 reiterated the conclusions of its subcommittee two years before: a technical survey of Scottish water resources and supplies should be undertaken at once and comprehensive enquiry should then be held into the whole question of water supplies. Three members of the committee were not content to leave matters there. In a note of reservation they took the view that water and drainage (i.e., sewerage) were the foundation of all sanitary measures and should be administered in areas large enough in both population and rateable value to secure the best results. This aim could be achieved only be establishing schemes that were sufficiently large to secure the supervision of skilled experts. Their views merit quotation in the light of subsequent events (DHS, 1936a):

> Our colleagues are not prepared to recommend any recasting of local authority functions applicable to water and drainage so long as there exists the possibility of combination and co-operation. But to rely on co-operation is to ignore the lessons of history in local government. While powers of combination and co-operation have existed for many years and the need is, or should be, self-evident, it is only in a limited number of cases and for special reasons that these have been exercised. Pressure by the central department is possible and on occasion may be effective, but where, as in the provision of water, the whole country would require to be covered by a series of joint boards, the necessary coersion of a vast number of small authorities would throw an intolerable burden on the central department. To hand over to a government department powers which ought to have been exercised by the legislature is to render those powers largely ineffective.
> (DHS, 1936a, p. 363-364)

In short, the three members saw the solution as a system of regional water authorities, to be established by specific legislation. The report of the full committee (DHS, 1936a) had recognised that the outstanding difficulty was

> ... that some of the town and county councils are unable out of their own resources to provide economically and efficiently for water supplies and drainage, hospitals, specialist medical and other services that in modern conditions, require large administrative units.

But it then added,

... to plan these services on a regional basis ... does not necessarily involve departure from the present local government structure ... the existing organisation ... provides for creating large areas, by co-operative action among the authorities to meet whatever need may arise.
(DHS, 1936a, p. 293-294)

The Committee did not think that the powers of the DHS should be strengthened except when it could be demonstrated that separate action by local authorities would involve greater expense and lower standards of services. It recommended that, in such circumstances, DHS should have power to demand the production of plans for the provision of services on a joint basis, and that if these still appeared the most reasonable approach after a public enquiry, they should be enforced by an Order.

The solution to existing and likely future demands was clear: regional co-operation would alleviate immediate problems through the pooling of surpluses and could do much to ease the burden of furnishing new supplies. Moreover, many rural areas were chronically underfinanced and a share of a comprehensive regional scheme would satisfy their needs more efficiently than any policy of independent action. Both problems and solutions were clear. Less obvious was the manner in which DHS could secure the recommended solution. As an essentially supervisory agency it had little power to intervene. Until it was given money to extend its area of influence or coercive powers, further delay in their adoption of a regional approach was inevitable.

PLANNING FOR POST-WAR DEVELOPMENT

It was not until 1943 that the technical survey of all Scottish water supplies, which the Committee had advocated, was undertaken. It was completed within two years and full information was thus available to co-ordinate a wide variety of measures stemming from programmes of planning for post-war economic and social development in Scotland.

Two reports were highly influential in this process. The Report of the Royal Commission of the Distribution of the Industrial Population (the Barlow Report, 1940) laid the foundation stone for post-war town and country planning and attempted to rectify imbalances in opportunities for employment between one part of the country and another. The Beveridge Report on Social and Allied Services (1942) urged that public responsibility should be admitted for securing to all citizens, regardless of where they lived, an important part of their fundamental needs for education, health services and housing, and an insurance system that provided some insulation from fear of sickness, accident or old age, in short "the welfare state". An increased interest of central government in planning, industrial development in the regions, housing and health would thus mean a change from central supervision over local powers to central direction of the activities of local authorities in their various functions, including water supplies.

Although the comprehensive survey of water resources by DHS was not completed until 1946 it was clear by the time of the publication of a White Paper on National Water Policy in 1944 (concerned with the whole of the United Kingdom) that many of the smaller systems of water supply in rural areas were inadequate at any time and severely lacking in dry spells, and that there was a distinct poverty of adequate treatment (DHS, 1944). It was in fact a reiteration of the problems which had been addressed by the Rural Water Supplies Act of 1934. Ten years had thus passed with relatively little action.

The Rural Water Supplies and Sewerage (Scotland) Act 1944

New legislation, the Rural Water Supplies and Sewerage (Scotland) Act 1944, made available £6.4 million in grants. These were intended to provide in the rural areas the basic services that were taken for granted by town dwellers. The need was clear. A survey undertaken in 1936 by the Scottish Housing Advisory Committee of three typical rural parishes had revealed that 67 per cent of the houses had no internal supply of water and in two-fifths of these water had to be carried more than 25 yards (Sherriff,1944). DHS announced that it wished to allocate the money only after all schemes had been submitted and it was some time before any schemes got underway (Sherriff, 1944). On the completion of the survey of water resources, 25 reports outlining potential regional schemes were issued to the appropriate authorities, with the suggestion that they should be considered as the basis of plans. Nineteen of the reports covered the supply of large areas and envisaged joint action by several authorities. By 1951, almost every county council had schemes for the improvement of rural services under consideration (Burns, 1952). The four hundred applications for grants that had been submitted would have cost some £25 million, qualifying for grant aid of £8 million. In view of the demand and of the need revealed by the DHS survey, the total amount made available was increased to £20 million in 1949, £30 million in 1955, £45 million in 1963 and £60 million in 1969, though with an increasing emphasis on sewerage and sewage treatment. Post-war shortages of materials and labour prevented an immediate start but by 1952, schemes were under way covering approximately one-third of the total area requiring general piped supplies. By 1966, 95 per cent of the population were in receipt of a piped water supply and by 1971, 98 per cent. The problem had been solved as, one after another, schemes went forward with DHS regional reports as the basis for action.

The intention set out in the White Paper, a National Water Policy (DHS, 1944), was to outline ways of ensuring that all future needs for water could be met. Sources of water in Scotland were more than ample: the problem was one of organisation and distribution. Three needs were identified: to extend piped supplies to all; to secure the most economical and effective use of existing resources; and to build up an accurate body of information. Measures to satisfy the first and the last were already under way, although a mechanism would be required to ensure that the data held at the centre were kept fully up-to-date in a routine manner.

Action was urgently required to secure the most effective use of existing water supplies. Although several water undertakings did transfer water across their official boundaries and a number had merged to form six ad hoc water boards, there had been a general lack of co-operation. The government was convinced that the multiplicity of small undertakings could provide more water more efficiently and more economically if they were to combine for the purpose. Echoing the conclusions of the Committee on Scottish Health Services of 1936, the government thought it preferable that combined action should occur by agreement; but where it did not occur the Secretary of State should be empowered to bring it about as a last resort, in the "public interest".

The Water (Scotland) Act 1946

The first major move towards the adoption of a more comprehensive approach to water management in Scotland came in the Water (Scotland) Act 1946. It had three major effects. Firstly, it synthesised the jumble of previous legislation into a single code. Secondly, it removed the need for local legislation and so saved Parliamentary time and avoided Parliamentary bottlenecks. Thirdly, it brought about institutional innovations, particularly concerning the role of central government, that were intended to give effect to the National Water Policy.

Under the Act, the Secretary of State was given several new duties as political and legal head of DHS. Firstly he was to promote the conservation of water resources, in the sense of making the most effective use of them or of promoting their optimal development. Secondly, he was to collect and publish statistics, and DHS was given a statutory right to information on water resources planning in Scotland. Thirdly, to broaden the range of consultation, the Secretary of State was to appoint an advisory committee, subsequently entitled the Scottish Water Advisory Committee (SWAC).

Central government was also given substantial powers to intervene as the terms and conditions of all acquisitions of water rights by local authorities were to be referred to DHS and all proposals involving capital expenditure had to be approved by the Secretary of State before they would be implemented. Finally, the Secretary of State was to act as a court of appeal in the event of a dispute between a water authority and other interests in the watercourse: he had to be satisfied that arrangements had been made to ensure that an 'adequate' flow remained in streams.

Central government was thus to promote an efficient water service by monitoring the performance of local authorities and by vetting their proposals. The role was outlined largely in passive terms and there was little power to initiate specific actions, at least overtly. The traditional reliance on persuasion was therefore destined to continue and local authorities were to anticipate the reserve powers of central government by presenting joint schemes as their first choice where these were appropriate. The government apparently felt that grants now available under the Rural Water Supplies Act and under legislation concerning industrial development would help in this process, even though such legislation did not require joint schemes as a condition of financial assistance.

The Water (Scotland) Act 1949

Although the 1946 Act made provision for metering and charging for industrial consumption, arrangements for domestic water charges were codified later. These had been considered by a Committee on Water Rating which reported (DHS, 1946) that the greater part of the population lived in areas of the 63 authorities which were supplying water under local Acts and that there were seven different systems of rating in operation.

The Committee examined several alternatives to the use of rates as a basis for water charges. One was to make a charge according to the number of water-using fittings in each dwelling, but the Committee believed that such a system would discourage both the use of water in personal hygiene and the introduction of WCs and baths into houses at present lacking them. Such a consequence would be unacceptable in the light of the Scottish Housing Advisory Committee's estimate (1944) that 405,000 houses out of a total of 1,300,000 (31 per cent) had either no independent water closets, no water closet at all or no sanitary conveniences of any description (DHS, 1946). Interestingly, however, the Committee offered no specific evidence of major outbreaks of waterborne disease which resulted from this circumstance.

Secondly, consumption could be metered and each consumer charged according to the quantity used. But the Committee did not favour this either, saying "while there may be a case for it in a country not so rich in water resources as Scotland, there is none where the problem of water supply is largely one of organisation and distribution" (DHS, 1946, p. 7). In addition, a system of metering would have the disadvantage of incurring high initial capital expenditure on meters and would involve the employment of a large staff of administrators and inspectors.

The Committee recognised that valuation for rating purposes was not an accurate index of occupancy or personal habits, but thought that a rating system would operate quite fairly on average. Indeed, it had received evidence (unpublished and unspecified) which indicated that the product of a metered charge would not materially differ from the sum actually charged under the rating system.

The Committee wished to see the adoption of a separate domestic water rate, presented to householders in a distinctive manner, because the inclusion of water charges in the general rate tended to draw public attention away from the value for money they were receiving from the service. The water rate had sometimes been regarded as prohibitively high because it was measured against the level of rate for public services as a whole (which at the time was relatively low) and not against the value of the water service to the individual.

The Water (Scotland) Act 1949 gave effect to the Committee's recommendations by instituting a uniform system of domestic water rates. It also abolished the contentious system of Special Water Districts for rating in rural areas. Henceforth, the same level of water rates would apply over whole counties, an approach that gave considerable scope for subsidising one part of an area by another. This adoption of a uniform domestic water rate within each local authority area had several important repercussions. Firstly, the financing of the water service was inherently associated with the financing of local government services as a whole, where the practice of raising capital by public borrowing means that over half of the expenditure on capital-intensive services such as water and sewerage relates to debt and interest payments. Faced with increases in current costs elsewhere, local authorities might be unwilling to commit themselves to further forced expenditure on a service which appears to have relatively low priority in political terms. New schemes would also have quite a recognisable and significant effect on the local level of rates, a factor which might prevent their being undertaken until they were necessary.

Secondly, the element of redistribution of wealth involved in the rating system inevitably characterises decisions on water management as 'political' and ultimate control of the service should hence rest in the hands of elected representatives of those whose wealth is being redistributed. This effect of spreading costs worked against the National Water Policy of promoting amalgamations for the more satisfactory development of new services and the better use of existing surpluses. In almost any proposed combination some authorities would be disadvantaged because the process would involve their taking a share of other people's costs and hence an increased rate burden with no visible benefit to their ratepayers.

THE NEW ACTS IN OPERATION

Such problems did not, however, assume a critical significance for another fifteen years. In the meantime, with an institutional framework that had been thoroughly overhauled, DHS seemed optimistic as local authorities got down to the task of examining their suggested schemes of improvement. This mood of optimism was reflected in DHS's annual report for 1948:

> With the augmentation of technical staffs of local authorities, the bringing up-to-date of the water code by the 1946 Act, the promise of grants for improving water supplies in the development areas and in rural areas and the general encouragement given by the Department following their engineering survey in 1943-45 to the planning of schemes on a wider and more comprehensive basis, Scottish local authorities now have before them a programme of £60 to £65 million about two-thirds of which relates to rural areas. This will keep the authorities busily engaged on these services for the next fifteen to twenty years.
> (DHS, 1948, p. 63)

It seemed that the Secretary of State's coercive powers would not be necessary.

It was not long, however, before clouds appeared on the horizon. Shortage of materials and labour in the post-war years inevitably restricted progress with several schemes. The Annual Report of DHS for 1950 announced that it had been necessary to tell a number of local authorities that work on particular schemes must be deferred until a place could be found for them in the construction programme.

> It has been demonstrated on a number of occasions that overauthorisation of work leads to longer delivery periods of essential materials, the under-manning of contracts, especially where the work is in remote areas, and the general slowing down of work of a similar type.
> (DHS, 1950a, p. 79)

As soon as these shortages had eased, progress began to be affected by a series of retrenchments in public expenditure that characterised the so-called 'stop-go' policy of central government in the 1950s. The Annual Report of DHS for 1957 announced that general restrictions on new loans had come into operation in February 1956 and that no consents to borrow capital had since been granted for any new scheme or expansion of an existing scheme except where considerations of health, safety or other vital interest had made deferment impracticable (DHS, 1957). Some 86 schemes of water supply were deferred in 1956 and 1957 and, in this atmosphere of austerity, it is likely that goodwill amongst at least some local authorities evaporated.

In the late 1950s, the prospect of major new developments arriving in line with the rapid restructuring of the economy and requiring large quantities of water within a very short timespan apparently set DHS engineers thinking along the lines of a very large regional scheme for the whole of the Central Belt. Any scheme would have to satisfy four requirements: the source should have a very large potential yield; it should be capable of development in stages; its capital cost should be as low as possible since no one could possibly predict how rapidly demand would increase; and it should be as near as possible to the Central Belt (Cormie, 1970).

A working party was formed in 1960 from those local authorities with a potential interest in water from such a scheme and it examined various possibilities. In August 1961, however, consulting engineers reported that a scheme based on pumping water from Loch Lomond was sound on both engineering and economic grounds. A Loch Lomond Committee was therefore formed to secure a Water Order, but this was to be no repetition of the co-operation which got post-war regional schemes underway. To understand what happened it is necessary to return to government policy on industry and economic development.

RE-ASSESSMENT IN THE 1960s

There was a pronounced swing against the government in Scotland and Northern England in the general election of 1959 although its Parliamentary majority had been increased. The reason was clear: whilst the country as a whole was enjoying a boom, the economy of these two regions was in decline, with unemployment in Scotland in 1960 more than twice the national average. As a result, regional economic policy in the 1960s was accorded a much higher priority by government than it had been in the 1950s, and had a particular focus upon the problems of Scotland and the North East Attention was directed at ways of promoting regional expansion and at the contributions which the regions might make to a higher rate of national growth.

The importance of creating an environment favourable to growth and the role which New Towns, urban renewal and a revitalized infrastructure of industrial services could play in this respect began to receive recognition. Very little of this analysis was original. The importance of promoting sound economic growth based on areas capable of expansion and of the links between regional (economic) and physical (town and country) planning had been emphasized by the Barlow Report some twenty years before. Specifically with regard to Scotland, the Cairncross Report of 1952 had assessed the problems and made recommendations (Scottish Council, 1952). Faced by the end of the decade with a worsening economic and employment situation, the Secretary of State suggested in November 1959 that the Scottish Council (Development and Industry) should establish an enquiry into the Scottish economy, with the full co-operation of government departments.

This enquiry was mounted by the Toothill Committee which reported in 1961 and made a wide variety of recommendations (Scottish Council, 1961). Of particular interest is its view of the role of central government, since it recommended that a new department should be created within the Scottish Office to bring together the industrial and planning functions of existing departments. The Scottish Development Department (SDD) was accordingly established in June 1962 to take over the duties of the Scottish Home Department with regard to industry and development (electricity, roads and local government), together with those of the Department of Health for Scotland relating to housing, town and country planning, water and sewerage. A new phase of government intervention in local affairs was about to begin.

The new department wasted no time and within a year produced the first regional economic plan of the 1960s, 'Central Scotland - A Programme for Development and Growth' (SDD, 1963a). Figure 3.1 shows its principal components. Few of these were new initiatives. What was new was their presentation as a single package and the commitment of the government to foster their implementation in a concerted manner. Of particular significance for water supply were two requirements: the provision of infrastructure services, and the other of investment grants and other financial incentives required by industries that the government wished to attract to the region. Second, five growth areas had been chosen as potentially the best in locations for industrial expansion and as foci of services. These were the new towns of East Kilbride, Cumbernauld and Livingston; the new town of Glenrothes; the Irvine District of North Ayrshire; the Grangemount-Falkirk area; and the Vale of Leven District of Dunbartonshire. Abundance of water could be a particular attraction to certain types of industry and in this respect the plan contained measures to ensure that Central Scotland's foreseeable needs were met. A water supply scheme focussed upon Loch Lomond was to provide up to 100 million gallons per day and would serve all the major developments mentioned in the paper except those in Fife and Ayrshire. Fife's needs could be met from other sources and Ayrshire's would be provided by a scheme based on Loch Bradan, which had been under discussion since 1956 and is described later in this chapter.

The new department also turned its attention quickly to problems of local co-ordination and co-operation. The structure of local government was examined to see how far

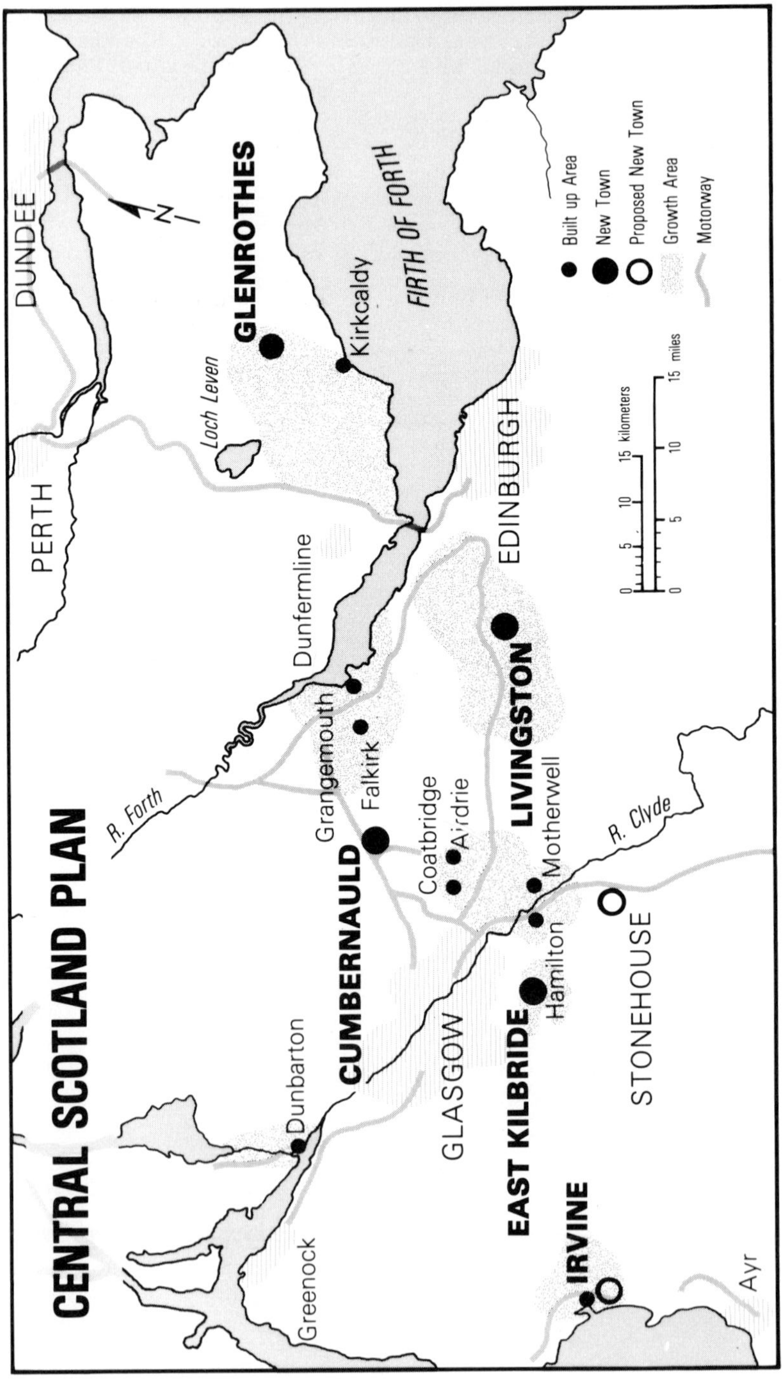

CENTRAL SCOTLAND PLAN
DUNDEE
PERTH
N
GLENROTHES
Loch Leven
Kirkcaldy
FIRTH OF FORTH
Dunfermline
EDINBURGH
R. Forth
Grangemouth
Falkirk
CUMBERNAULD
Coatbridge
Airdrie
LIVINGSTON
Motherwell
R. Clyde
Dunbarton
Greenock
GLASGOW
EAST KILBRIDE
Hamilton
STONEHOUSE
IRVINE
Ayr
0 5 10 15 kilometers
0 5 10 15 miles
Built up Area
New Town
Proposed New Town
Growth Area
Motorway

Figure 3.1

it matched the needs of "the expanding Scotland of today and tomorrow" and SDD's views were published in 1963 in a second White Paper (SDD, 1963b). Since the last reorganisation of local government in 1929, economic development had become critically dependent upon local authorities providing the necessary infrastructure, notably housing, roads, water and sewerage services. As a consequence, a new type of administrative framework was required. Among the suggestions was a two-tier form of local government: those services which would benefit most from the administration of a large area should become the responsibility of large regional authorities (the upper tier), while those services for which more local control was thought appropriate should be allocated to a second, lower, tier of authorities, perhaps created through the amalgamation of burghs with their surrounding rural areas. Water supplies should be the responsibility of the regional authorities, as should sewerage and river purification.

THE SCOTTISH WATER ADVISORY COMMITTEE'S INVESTIGATION

The White Paper on local government was prepared for consultation rather than as a firm plan for immediate action. In the meantime, however, particular services such as water supplies required more specific proposals. A policy for water had existed since 1944: the time had now come to develop a programme of action, at least with regard to Central Scotland. The new Scottish Development Department was awaiting the views of the Scottish Water Advisory Committee (SWAC) which had been commissioned by DHS in January 1962 to examine local administrative control over the development and distribution of water supplies in central Scotland and:

> ... how far it might be desirable to draw together local water authorities in the area, with a view to facilitating measures for securing an efficient and economic supply of water adequate for all purposes, throughout the area. (SWAC, 1963, p. vii)

This was the thorough enquiry into the legal and administrative aspects of water supply that the Committee on Scottish Health Services had urged some twenty-five years before.

In effect, SWAC was to hear options and assess the viability of a long-standing policy. Its members were largely professional engineers in the public service drawn from a wide range of authorities, but all had one thing in common: an understanding of what was possible in the world of central government relations. As with the Toothill Committee, SWAC was assisted in its task by permanent officials of DHS (subsequently SDD).

SWAC was conciliatory about the existing pattern of development and distribution in Central Scotland. It recognised that the primary duty of the individual authorities had been to supply water to users in their own areas, and that separate development of sources may well have been a rational course of action when viewed in the light of then-existing circumstances. Times had changed, however, and a new approach to water supply seemed essential. Not only was there a wasteful duplication in some areas but others still lacked the necessary financial resources to provide adequate supplies. The most urgent action was required in Central Scotland and the initial attention was focused mainly upon that region (SWAC, 1963).

SWAC welcomed the proposal to embark on the Loch Lomond scheme which it considered, "to be conceived on the bold and imaginative lines necessary to match the challenge of the rising demand for water". Indeed, in its view, "the aim must be to align the maximum possible support for the scheme ... to broaden the back that must bear the heavy expenditure involved". It would be much easier to achieve the necessary co-ordination if there were fewer and larger water authorities. SWAC believed that the Loch Lomond scheme made a major reorganisation of the administrative structure a matter of compelling urgency.

There were too many water authorities which were too small and too weak to fulfil the role expected of them. The Secretary of State had made it clear that a solution was expected that involved some sort of regional board of which there were two possible types: bulk supply boards, which would furnish water to existing local water authorities for distribution, and "source-to-tap" boards responsible for both supply and distribution. There were Scottish precedents for both, but the former had already been rejected in England and Wales where a circular of July 1958 had announced:

> ... the Minister is of the opinion that a general system of bulk supply boards, with distribution in the hands of existing water undertakers, would be wasteful of manpower and resources and that, in order to meet the overriding requirements of an efficient and economical water organisation, unified control over supply and distribution is essential.
> (SWAC, 1963, p. 18)

SWAC endorsed this view, arguing that a system of bulk supply boards would increase rather than reduce the number of water authorities. There would still be the danger of supplies that were not immediately required being reserved by a distribution authority in recognition of the capital contribution it had made to making them available. Another authority's urgent needs might therefore be denied. Neither greater co-ordination nor flexibility would necessarily follow the introduction of such a system.

The Institution of Water Engineers favoured a single board for the whole of Central Scotland. Such an institution it argued, would ensure the best use of existing sources and facilitate rational forward planning. Day-to-day operations would require the delineation of six divisions which would reflect topography, be of a size most suitable for efficient routine maintenance and take no account of the existing boundaries of the local authorities. The Board would levy a single water rate (set at the same level for the whole of its area) and be appointed by the Secretary of State.

SWAC saw a merit of such a scheme from the engineering point of view but found it politically unacceptable and therefore impracticable. Firstly, the suggestion of replacing existing authorities with a single board consisting wholly or partly of non-elected members would be unacceptable from the democratic point of view, bearing in mind the historical role the local councils had played in developing supplies. Even if the membership consisted entirely of the nominees of local authorities, some of the seventy or so authorities would not be adequately represented in relation to their size and others would not be represented at all.

Secondly, although the Institution sought a uniform charge for water over the whole area because of the extent to which engineering solutions to problems of supply had been adversely affected by the differential impact of joint projects on local rates, SWAC was not convinced that, "in present circumstances" a uniform charge was a practicable proposition (SWAC, 1963, pp. 39-40).

After weighing the evidence submitted to it, SWAC was satisfied that a system of 'source-to-tap' boards would meet several objectives. A small number of authorities of this kind would provide an administrative system which would remove difficulties of inflexibility and lack of co-ordination enable existing sources to be pooled for the common good and facilitate the degree of co-operation between sizeable authorities that was necessary for the development of major sources, such as Loch Lomond. Such a system would also remove another weakness: water management was too often the part-time task of technical officers who were heavily burdened with a wide variety of other duties. Only nine of the 61 authorities in Central Scotland had a full-time engineer concerned with water supply.

Just as it was essential to secure unified control within each region, SWAC thought there should be some provision for co-ordination and co-operation between regions. For this purpose it recommended the creation of a strong central water development board, whose first major task would be to oversee the Loch Lomond project. SWAC attached particular importance to the need to give the development board the power to borrow in its own right so that it would operate without the hindrance of having to reserve supplies for the sole use of particular regional boards. It would then finance its debt by the sale of water to regional boards at a uniform charge. A system was required whereby the high initial costs of developing Loch Lomond would be recovered as and when the reserves made available were taken up.

The Institution of Water Engineers, the Convention of Royal Burghs and the Federation of British Industries had all suggested that there was a need for a central authority in the administrative structure of the water service, the task of which would be to promote co-ordination and resolve disputes between boards. SWAC reminded them, however, that this was the duty of the Secretary of State and was strongly of the opinion that his department should not only continue to settle disputes, "as an independent authority above the battle" but also continue to exercise a strong role in the overall co-ordination of the development of new sources (SWAC, 1963, p. 23).

Finally, in considering the areas to be covered by the proposed regional boards, SWAC had taken account not only of existing local government boundaries but also of physical features and technical considerations. In certain cases the latter made it necessary to depart from these boundaries. Six 'source-to-tap' water boards were recommended, the areas of which are shown in Figure 3.2.

SWAC submitted its report on Central Scotland in March 1963. The government's programme for development and growth was published in November of that year but had defined Central Scotland to include Ayrshire and Renfrewshire whilst SWAC had excluded them. Accordingly, SWAC was invited to expand its investigations to encompass these two areas.

WATER FOR CENTRAL SCOTLAND

Developments in Ayrshire illustrate the difficulty of advancing on a voluntary basis. The Irvine area of North Ayrshire had been designated a growth area and, as has already been noted, the Central Scotland plan stated that water could "be fully supplied from the Loch Bradan scheme on which consultations were taking place" (SDD, 1963a). This loch had also been suggested as a source of future supply by DHS in the post-war regional review. Ayrshire County Council had employed consultants in the early 1950s to confirm the viability of such a scheme but had been unable to reach agreement with other water authorities as to its full development. Loch Bradan had already been partially developed by Troon Burgh Council and was hydrologically connected to the catchments providing Ayr Burgh Council's water supply. In addition, the Irvine and District Water Board had hitherto shown little interest in a co-operative redevelopment of the source though the plan for Central Scotland had stated that, "an expanded water supply from this source will be essential to cater for the substantial growth of population envisaged in the Northern part of the county" (SDD, 1963, p. 21).

S.W.A.C. REGIONAL WATER BOUNDARIES - 1963
N
FIFE AND KINROSS-SHIRE
WEST LOTHIAN EDINBURGH AND MIDLOTHIAN
STIRLINGSHIRE AND CLACK MANNAN
LANARKSHIRE
GLASGOW AREA
DUNBARTONSHIRE
Perthshire
East Lothien
Berwickshire
Roxburghshire
Selkirkshire
Peeblesshire
Renfrewshire
Ayrshire
kilometers
miles

Figure 3.2

One month before the publication of this plan, the Scottish Office had convened a meeting of all the water authorities in Ayrshire in the hope of getting agreement on an early start to the regional scheme but there was no positive response. The Secretary of State accordingly called upon SWAC to investigate the possibility of establishing a regional board.

SWAC concluded that the Ayrshire case exhibited all the characteristics which had led that body to recommend the establishment of regional boards in the rest of Central Scotland (SWAC, 1964). The Committee had also been impressed by the need for prompt action in view of Ayrshire County Council's dependence on the development of Loch Bradan. It was SWAC's experience during the investigation that let it to believe that there was little hope of voluntary agreement on the scheme and that further negotiations would merely lead to still more frustrating delay. Resolute action by the Secretary of State was necessary. The Committee's view was that he should ask the authorities whether or not they would accept a regionalisation of the water service and, if not, he should make an Order to bring it about compulsorily SWAC, 1964, p. 1).

The Scottish Office convened a further meeting in July 1964 to elicit the response of local authorities to SWAC's report. No significant progress was made, however, so the Secretary of State published his intention in December 1964 of making an Order compulsorily regionalising the water service in Ayrshire and authorising the Loch Bradan scheme. Predictably, however, nine of the fifteen water authorities affected lodged formal objections and it was announced that a public inquiry would consider these objections. At the inquiry, which opened on June 17th, the Town Councils of Ayr, Prestwick, Girvan and Largs all objected that the Order would not give them a better water supply and would, even without the Loch Bradan scheme, result in substantially increasing water rates (Ayr Advertiser, 17/6/65 and 24/6/65).

The Recorder conducting the inquiry concluded that the Order would be of overall advantage to eleven of the fifteen authorities and secure for them a better supply. The remaining four would also derive some advantage from their inclusion in the scheme and would secure a better supply in the long term, although their current need in this respect had not been established (Munro, 1966).

Yet the public inquiry into regionalisation and the Loch Bradan scheme did not end matters. Ayr Burgh Council announced its intention of taking the issue to Parliament, following the lead of the City of Edinburgh which had pursued similar objections to the much larger Loch Lomond scheme.

ALTERNATIVES FOR THE LOTHIANS

Experience in the Lothians further strengthened SWAC's view that more radical action was necessary. The Loch Lomond scheme had likewise encountered difficulties, particularly over the supply of Livingston New Town, part of which lay in Midlothian whose county council had merged its water undertaking with that of Edinburgh in 1949 in response to a recommendation from DHS (Fig. 3.3). The rest was the responsibility of the West Lothian Water Board which had been created in the early 1950s, also in response to suggestions from DHS. In 1960, this board had obtained powers to proceed with a West Water scheme which had been the planned source of future supply for some time, but with the certainty of a major new vehicle plant being established at Bathgate, it became clear that this scheme could assure future supplies only until 1967, a mere two years after water would become available. The West Lothian Board accordingly began an urgent investigation of further sources of supply. Consultants recommended the use of the last undeveloped source in the Pentland Hills, but SDD drew the board's attention to the Loch Lomond scheme. It accordingly joined the Loch Lomond working party and eventually became the authority with the largest single interest in the scheme, with a reservation of 17 million gallons per day out of the proposed total supply of 100 million gallons per day.

Simultaneous involvement in two major new developments (the West Water and the Loch Lomond schemes) would more than double the board's water rates, which were already high. SWAC took the view that the board's dependence on the Loch Lomond scheme for future supplies and its very high water rate were clear evidence that it was too small to be a viable water undertaking. Topographic obstacles affecting the distribution of water made a combination with water authorities to the north and west impracticable and SWAC therefore recommended that the board should merge with water undertakings in Edinburgh and Midlothian (SWAC, 1963).

Edinburgh Corporation had also considered Loch Lomond as a potential source of future supply, but consultants had advised against participation, recommending instead a further development of the Upper Tweed basin, its traditional source of supply (Edinburgh Corporation Water Department, 1963a, 1963b). In view of SWAC's proposal that Edinburgh should become involved in the supply of West Lothian and, in particular, Livingston New Town, the consultants were asked to cost supplies from the Upper Tweed and compare them with supplies from Loch Lomond. They concluded that costs were similar but that, for a number of other reasons, the Upper Tweed would be the most appropriate source of supply; it would involve considerably less pumping than the Loch Lomond scheme and would have lower running costs, especially if the cost of electricity continued to rise. The Corporation was therefore unwilling to participate in the Loch Lomond scheme and the West Lothian Water Board was left without the support of its large and more affluent neighbour.

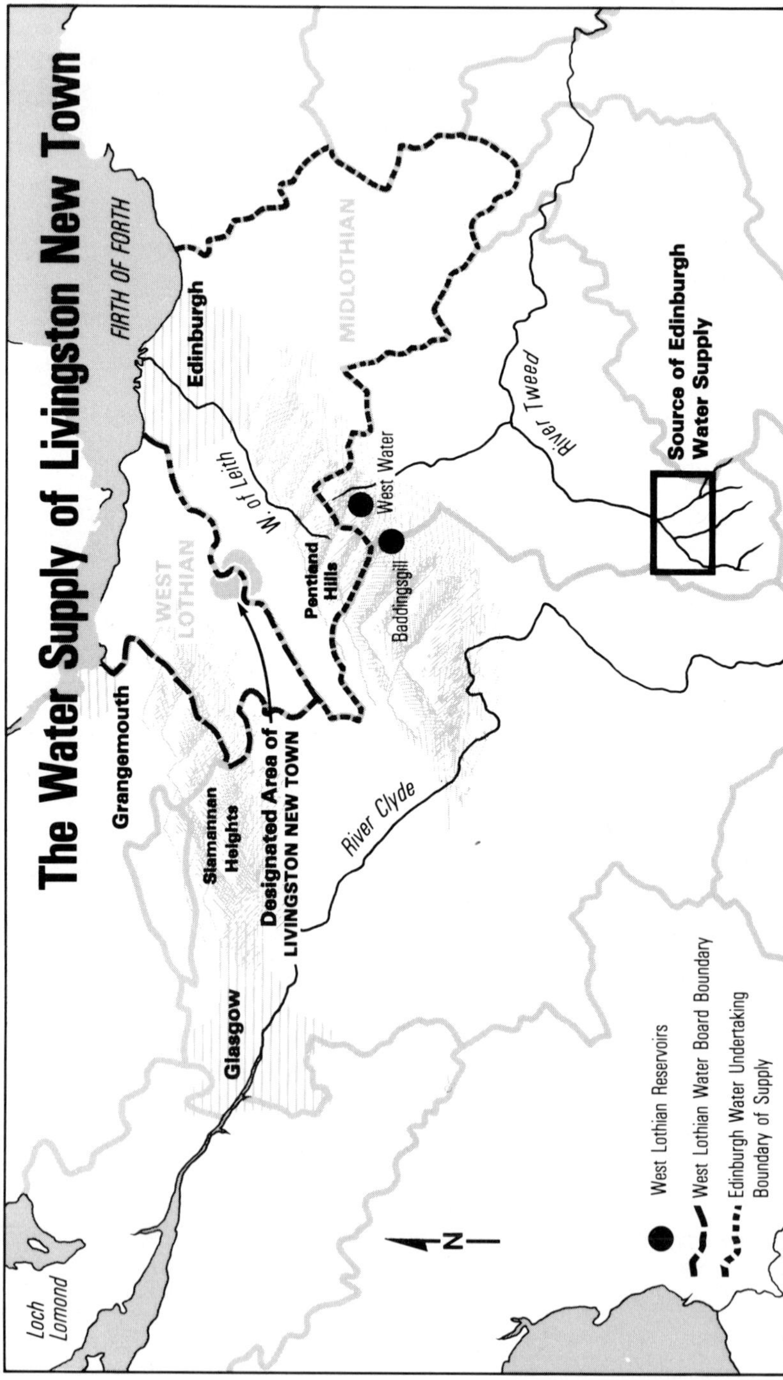

The Water Supply of Livingston New Town
FIRTH OF FORTH
Edinburgh
MIDLOTHIAN
WEST LOTHIAN
W. of Leith
Pentland Hills
West Water
Baddingsgill
River Tweed
Source of Edinburgh Water Supply
Grangemouth
Slamannan Heights
Designated Area of LIVINGSTON NEW TOWN
River Clyde
Glasgow
Loch Lomond
N
West Lothian Reservoirs
West Lothian Water Board Boundary
Edinburgh Water Undertaking Boundary of Supply

Figure 3.3

There were thus many parallels with the Loch Bradan case, with a single town standing out against the tide. A draft Edinburgh and West Lothian Water Board Amalgamation Order had in fact been published in 1964 but it was not implemented because of the resolute nature of Edinburgh's objections. The Scottish Office had apparently hoped to get the Loch Lomond scheme underway before tackling the delicate issue of Edinburgh's involvement with West Lothian and Livingston, but the Corporation, though not officially an affected party, persistently laid formal objections to the scheme, even to the extent of petitioning Parliament over the matter in 1966 (The Scotsman, 3/12/66).

IMPLEMENTING THE NEW POLICY

The 1964 general election brought a change of government and a new Secretary of State to the Scottish Office. A shift in priorities was expected, particularly with respect to Scotland and Northern England. Regional development was to be accorded particular emphasis. Within a year the government announced its "Plan for Expansion", a principal feature of which was the extension of the "strategy of growth areas" to the whole of Scotland (SDD, 1966). The provision of adequate water supplies was an essential ingredient in such plans. The previous Secretary of State had sent copies of SWAC's 1963 report on Central Scotland to all water authorities outside the central belt and had invited these authorities "to consider whether there was scope, in the general interest, for joining with other neighbouring authorities into units which can make more economic and flexible use of water resources than is possible, each by itself" (SWAC, 1966, p. 1). This invitation had evoked little response by November 1965 when the new Secretary of State asked SWAC to extend its study to those other areas.

SWAC began this further investigation in the full knowledge that the principle of its previous recommendations had been accepted in July 1963 by the government. In November 1964, the new government had also adopted as policy the system of "source-to-tap" regional boards and a water development board to administer the proposed Loch Lomond scheme. In these circumstances SWAC felt it unnecessary to review the arguments for and against regional boards. Its task was to ascertain if what was appropriate for the central belt would also be valid for the rest of the country; it would also take the opportunity to reconsider some of its earlier recommendations on regional combinations in the new context of a national perspective.

MOVING BEYOND THE CENTRAL BELT

The administrative problems of water supply were admittedly different outside the central belt. Of the 126 authorities,

93 supplied fewer than 1,000 people and this accounted for 86 per cent of the land surface (but only 28 per cent of population). Distance, unpopulated mountainous country and difficulties of communication had all to be taken into account. Nevertheless, SWAC felt that the number of water authorities was needlessly large and that most were too small and too weak to be able to play a viable role as separate authorities under modern conditions. The evidence gathered by the committee suggested that many authorities had failed to co-operate as they should have done. The differences between centre and periphery, however, were of degree and not of kind. The case for reorganisation rested on a need for viable units and the real question was not so much whether amalgamations should take place but rather which particular amalgamations would be best.

Two earlier recommendations required some modification in the light of the new national perspective, those relating to Renfrewshire and West Lothian. Since SWAC's report in 1963, consulting engineers had examined possibilities for the future supply of Renfrewshire and concluded that, in the long term (after 1990), the area would have to look north to Loch Lomond. In this light they had recommended that the county should be involved in the scheme from its inception. Other county authorities in the Lower Clyde would also be interested in the scheme and Renfrew County Council accordingly suggested an amalgamation with the proposed Dunbartonshire Regional Board. The county had felt that Glasgow should continue on its own, but other authorities believed that the city's financial resources would make a welcome contribution to any combination. Glasgow Corporation, whilst generally sympathetic to the concept of regionalisation, declined to comment on a wider grouping until it had more information on the technical and financial implications. SWAC, however, saw advantages in linking Dunbartonshire, Glasgow and Renfrewshire, and recommended the creation of a Lower Clyde Water Board embracing all three. With a population of 1.5 million it would not be out of scale with water management in England (SWAC, 1966).

When SWAC came to review the administration of water supplies in the Tweed Basin, the sensible solution seemed to be the amalgamation of all authorities with interests in the same sources. These included West Lothian, which had a traditional source of supply in the Pentland Hills, which it shared with Edinburgh and the Tweed authorities. SWAC, therefore, recommended the extension of the Edinburgh and West Lothian merger to include East Lothian (which already shared the Lammermuir catchment with Edinburgh) and the Tweed authorities, to create a single South-East of Scotland Water Board. The Committee felt that a strong regional grouping such as this would not only provide the viable water undertaking required by the new government's White Paper on development, which laid plans for expansion in the Boarders region, but also take some of the heat out of Edinburgh's view that the city's institutions for water management were being reorganised solely to underwrite expansion in West Lothian. Based on the evidence from various interests in the areas involved, SWAC proposed a national system of 13 boards as shown in Figure 3.4.

With respect to the proposed water development board, SWAC reiterated its view that the board should have its own borrowing powers and adopt a new system of charging for bulk supplies. Moreover, the board would superintend not only the Loch Lomond scheme but also any subsequent inter-regional schemes, although it was thought that these would only ever be required in Central Scotland (SWAC, 1966).

SWAC based its recommendations on evidence submitted by a wide range of interests, but particularly representatives of local and central government agencies, and professional associations. Two of the latter were especially concerned about the need for reform and argued strongly for an agency with a central co-ordinating role, outside the Scottish Office. In particular, the British Waterworks Association and the Institution of Water Engineers had a much wider vision of the role of the proposed water development authority. The Association suggested that an all-Scotland water development board might carry out the following functions:

1. Survey the water resources of each board area.

2. Assess demand in each area.

3. Determine the sources that would be developed to meet demand.

4. Plan programmes or work within each area.

5. Make available to each board specialised techniques and disciplines, which it might otherwise be unable to afford.

6. Provide specialised equipment, such as computing services.

7. Provide an engineering service for the more remote areas.

8. Advise the Secretary of State on the disposal of grant-aid for water.

9. Arrange mutual aid programmes both in men and equipment.

10. Work towards the adoption of standard equipment and practices.

11. Provide services for bacteriological and chemical analyses.

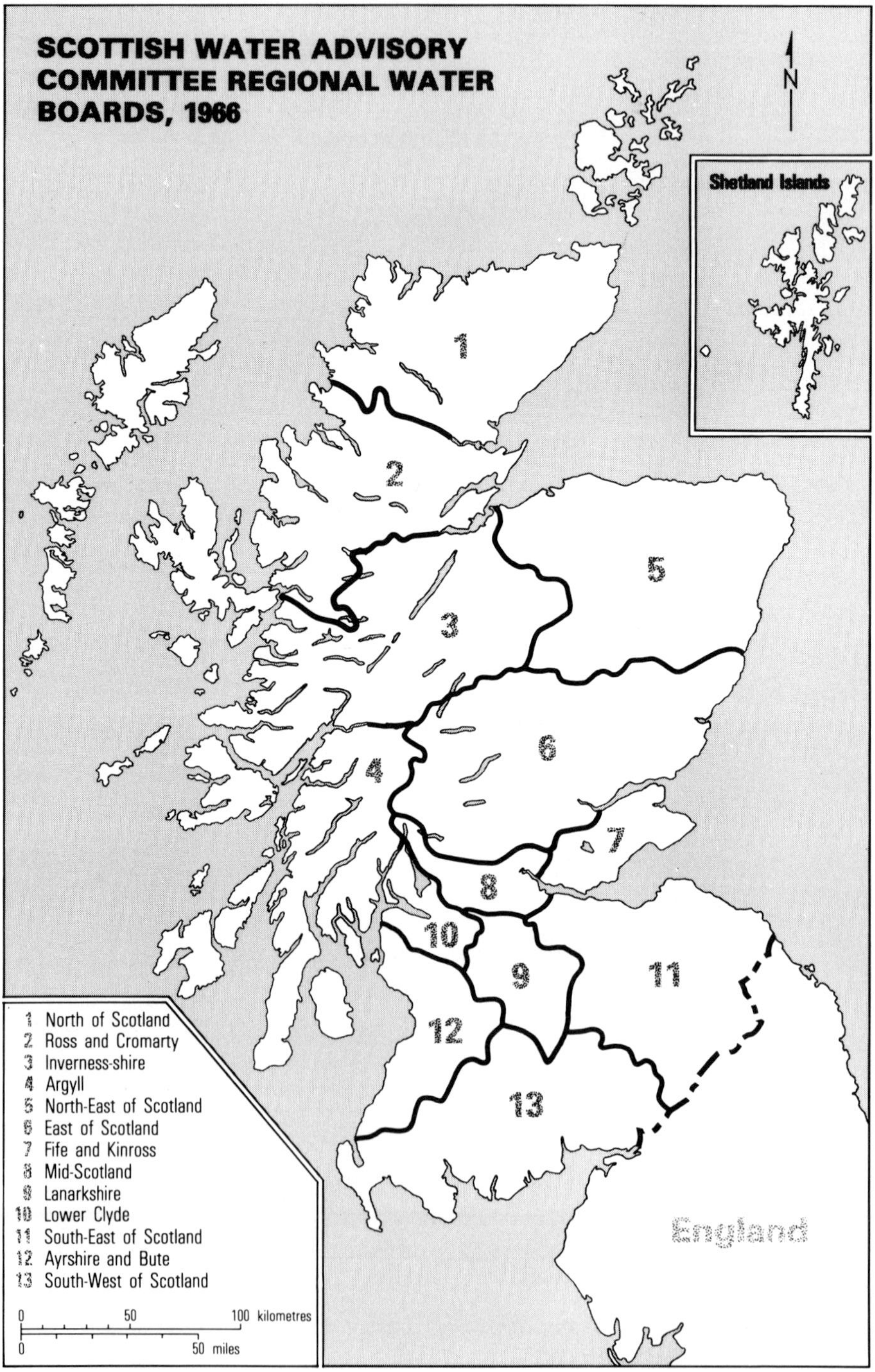
SCOTTISH WATER ADVISORY COMMITTEE REGIONAL WATER BOARDS, 1966
N
Shetland Islands
1
2
3
4
5
6
7
8
9
10
11
12
13
England
1 North of Scotland
2 Ross and Cromarty
3 Inverness-shire
4 Argyll
5 North-East of Scotland
6 East of Scotland
7 Fife and Kinross
8 Mid-Scotland
9 Lanarkshire
10 Lower Clyde
11 South-East of Scotland
12 Ayrshire and Bute
13 South-West of Scotland
0
50
100 kilometres
0
50 miles

Figure 3.4

In effect, the professionals were again arguing for a national water agency as they had done in representations three years earlier. All the functions are those of a headquarters organisation, with the regional boards merely being left with the task of day-to-day maintenance and, of course, of raising money to pay for new developments. On the other side, the Secretary of State would act as an agent of the Treasury and return to his pre-1946 role of administrative superintendent.

SWAC did not accept the recommendation of the professional groups. It rejected the idea of a central co-ordinating body, arguing that an additional executive tier of administration "would merely add a fifth wheel to the coach". If a development board were to be established, it should be an equal partner of the regional boards, not their master (SWAC, 1966).

SWAC recognised that its role was that of an advisory body. It was anxious to ensure, however, that its proposals stood a chance of being implemented. In this respect it saw the statutory procedure laid down in the 1946 Act as a potential stumbling block. In particular, it seemed likely that every regional amalgamation would be opposed on similar lines to experience in Ayrshire and each would require a public inquiry to be held. Even where objections were subsequently withdrawn, at least a year would pass before any Order was implemented. On the other hand, if objections were sustained and the case went to Parliament, a much longer period would elapse. The sheer weight of administrative work accompanying such a process would ensure that only a small number of Orders would be processed at any one time. Accordingly, it would be some years before the complete reorganisation came about and, in the interim, some authorities might well be reluctant to spend on urgently-needed improvements while the prospect of reorganisation hung over them. SWAC felt, therefore, that existing procedures could not be relied on to produce results with the speed that the situation demanded, notably in industrial areas that were developing rapidly (SWAC, 1966). Now that a national reorganisation of water management was being proposed, the Committee recommended that the whole matter might be considered in Parliament.

The Water (Scotland) Act 1967

SWAC's conclusions were published in September 1966. The draft Orders to rationalise water management in Ayrshire and to establish a Loch Lomond Water Board were both bogged down in procedures for dealing with objections, and the government's strategy for economic development seemed threatened. Accordingly, the Secretary of State for Scotland introduced a Water (Scotland) Bill to reorganise the water service in Scotland into the House of Commons in January 1967.

The Bill was presented to the Scottish Grand Committee with the following rationale (Hansard, 1967, cols. 6-7). The previous Secretary of State, when accepting the recommendations SWAC had made for Central Scotland in 1963, had hoped that the regrouping of authorities could be achieved voluntarily. This optimism was unfounded. In every case some local authorities had blocked amalgamation by insisting on putting the interest of their own rate payers before the wider interest of the economic growth and prosperity of Scotland. Compulsory amalgamation Orders had been published in 1964 but subsequent experience had demonstrated that it would take years to secure amalgamation by those means. The 1963 plan for Central Scotland had envisaged that the Loch Lomond scheme would already be in operation. In the Secretary of State's words, "We cannot afford to wait any longer. The need to develop supplies for new industry and new communities is so immediate that changes proposed in the Bill must be made now." No further steps would be taken to bring into effect any of the Orders for regionalisation that had been published in draft. An existing draft Loch Lomond Order should, however, go ahead immediately because of the urgent need for water in Central Scotland. This kind of action had to be taken, suggested the Secretary of State, if "we are to face our responsibilities as legislators and mean what we say about economic growth in Scotland".

In contrast to the 1946 Act, the 1967 Water (Scotland) Act was brief and single purpose. It empowered the Secretary of State to establish as soon as possible the regional water boards specified in a Schedule of the Act which were those recommended by SWAC the previous year. It also dealt with the transfer of assets, established a Central Scotland Water Board to develop inter-regional sources of water supply and allowed for subsequent modifications of the initial pattern of boards and the creation of further water development boards should that be necessary.

The new boards would have a duty to consult each other, to collaborate on matters of common interest and to give as early notice as possible to other boards of any investigation of a potential new source. The Secretary of State was empowered to determine, by Order, how constituent local authorities should be represented on the new boards. In carrying out this responsibility, he was to pay due regard to the rateable value and population of the area covered by each constituent council. The membership of the development board would be appointed from and by the constituent regional boards.

REORGANISATION IN PRACTICE

The thirteen boards created under the 1967 Act survived until 1975 when the water service was further reorganised and re-incorporated into the new structure of local government. It is fair to say that the period 1968-1975 appears to be regarded by water engineers in Scotland as their halcyon days, principally because they were given the opportunity to show what they could do in modernising the approach to water management. Before 1968, many water authorities simply had insufficient financial resources and water had a low priority among other diverse demands on local authority budgets. Except in the immediate aftermath of shortages, there were no votes to be gained from expenditure on the water service and sometimes not even then: when the Fife and Kinross Water Board took over from 19 local authorities there was only a two-three per cent margin of supply over demand, which was growing at approximately 2.5 per cent per annum.

New boards did not necessarily mean new thinking and at least one director was unable to convince his board of the need to plan ahead or to institute systems of waste prevention. In general, the first and principal task for the boards was to make a better use of what they already had and then progressively to integrate small supply systems. Nevertheless, a review undertaken in 1972 found that, in the three and a half years since the boards had been established, there had been far-reaching changes: new and improved services had been provided, distribution systems had been rationalised and maintenance and management had improved generally. It was the view of many officials that the service was at once modern and efficient and that the policy recommended by SWAC in 1966 had proved highly successful in practice (SDD, 1972).

A major element in this success had been the universal adoption of the 'source-to-tap' principle, which recognised that the major difficulties experienced in the past were rooted in administrative barriers. One of the principal benefits was the opportunity the new system afforded of providing additional supplies to one area by using spare capacity held in another, simply by integrating the two distribution systems so that surpluses might be transferred, an easy task if there were no opposing legal and financial interests. The larger the area under common administrative control the greater were the possibilities for such inter-area transfer.

Larger units of administration are advantageous not only for this reason but also because they make decision-making on developing new sources of supply internal to the organisation, thus avoiding the difficulties that arise from the adoption of inflexible postures by one party against another for reasons other than providing the most efficient service. Water boards met specifically to consider the needs of the service and members had been encouraged by the Secretary of State to think in terms of the needs of the region as a whole. In such a context, professional staff apparently found it much easier to promote good management.

The sequence of events described in this chapter has concentrated on only one aspect of water management. Nor, apart from the possibility of inter-regional transfers, did it involve major innovations, though its final stage, implementation of the Water (Scotland) Act, did lead to greater professionalism (if not a wider range of professionals). It also witnessed major changes in the size of the unit of management, culminating in the adoption of boards which correspond broadly with river basins or large parts of them, as the "source-to-tap" principle implies. In many ways, it represents the apogee of single-purpose management. The new system did not, however, last long. The reform of local government in 1973 was soon to reduce the number of possible constituent authorities, thus making it possible once again to maintain the principle that only those who were responsible for the provision of finance and who faced the political repercussions of expenditure should have the final say, not only so that they could supervise how money was spent but also so that expenditure could be properly co-ordinated with other aspects of spending by local authorities.

4

Broadening the perspective in the management of water quality

As with water supply, the basic problems of managing water quality in Scotland were defined in the 1930s, but were not fully tackled for thirty years or more. Here, too, a lack of professionalism in staffing and the fragmented nature of the local authorities, which have throughout been responsible for sewerage and sewage treatment, are important explanatory themes, as is contemporary practice in England and Wales. The distinctive feature, however, is the creation of the River Purification Boards and it is therefore appropriate to begin with the Scottish Advisory Committee on River Pollution Prevention, whose work has already been introduced in Chapter Two. Changes in legislation, policies and administrative structures of relevance to the management of river water quality are then considered.

THE SCOTTISH ADVISORY COMMITTEE ON RIVER POLLUTION PREVENTION

The evidence gathered by the Scottish Advisory Committee on River Pollution Prevention (ACRPP), considered in Chapter Two, convinced it that a drastic change in administrative arrangements for the control of water quality was necessary and in 1936 it published a report outlining the alterations to the law that it considered desirable (DHS, 1936b). The Committee concluded that:

> After fully reviewing the whole situation we feel the time has come when the problem of rivers pollution prevention should be dealt with on a much broader basis than hitherto, and that, wherever practicable, comprehensive schemes should be adopted whereby large sections of the country would be brought within the scope of these schemes. Convinced that the solution of the problem lies in the administration of the law by a joint body and that, ... this end would not be attained by joint committees as can be constituted under the existing law, we have given careful consideration to the constitution of an alternative body. In our opinion, effective action could be taken by a River Board representative of all

local authorities within the area of the watershed, together with representatives of the following interests - industry, land, fishing and navigation. The sole duty of the Board would be to ensure that the rivers within its area were not poluted, and experience has shown that a comprehensive or composite body of this nature, having one set purpose, is more disposed to see that the work entrusted to it is actively and efficiently undertaken.

Present practice was not only ineffective in many places, it was also unfair. Some local authorities had spent very large sums of public money purifying the sewage from their districts while others on the same river had done little or nothing, thus nullifying the efforts of the former; similarly, some industrialists had recognised their obligations while others had not. As a consequence, many industrial users or farmers were unable to use the polluted water as it came to them, while fishing interests, including the commercially-important salmon fisheries, had been seriously prejudiced and, in some places, completely destroyed.

The Committee submitted a ten-point plan (DHS, 1936b, pp. 11-12). Specifically, it recommended that:

1. DHS should define groups of river basins for the purpose of exercising control over pollution, including tidal waters.

2. A River Board should be established for each of these areas.

3. The membership of River Boards would reflect all interests in the rivers of the area; a majority of members would represent the local authorities of the area.

4. DHS should prescribe national standards for water quality; these to be modified by individual Boards to meet local conditions as was thought appropriate.

5. All producers of discharges to rivers would have a legal duty to observe the standards laid down.

6. Each local authority should be required to prepare a scheme for ensuring that the rivers of its area were not polluted. The authorities should have a statutory duty to consult each other in the preparation of such plans with a view to cooperative action.

7. The River Board would endeavour to co-ordinate the content and implementation of such plans to ensure that the most efficient and economical scheme would be devised for the area.

8. River Boards should appoint 'River Inspectors' and other professional staff to superintend the implementation of the law and supervise the completion of necessary remedial work by the River Board in the event of a failure on the part of an offender to carry this out.

9. The expenses of each River Board would be met by requisitioning the local authorities within its area.

10. Each River Board should publish an annual report outlining the present state and the achievements of each year.

The ACRPP also wished to see the application to Scotland of proposals which were about to be enacted in the Public Health (Drainage of Trade Premises) Act of 1937 for England and Wales. Briefly, these sought to impose upon local authorities the duty to receive trade waste waters into public sewers, subject to certain safeguards, and to confer on traders the right to demand a connection to public sewers (thus establishing a national system of rights and duties which would replace a host of local agreements and conditions).

These views and proposals produced no immediate response. They were NOT adopted in the National Water Policy of 1944 (as it applied to Scotland) (DHS,1944). The ACRPP's report was acknowledged but the government was not satisfied that the time was right for a move in this direction in the light of the extent to which sewerage and sewage treatment was known to require improvement, especially in rural areas (ibid., p. 23):

> The Government are satisfied that improvements with regard to the administration of the River Pollution Prevention Acts in Scotland are necessary, but having in view the different conditions prevailing there, it has yet to be determined whether these improvements can best be achieved by the setting up of river boards or in some other way. The Secretary of State accordingly proposes to open discussions on this question with the Association of Local Authorities and the other interests concerned at an early date.

The traditional means of dealing with water problems where the necessary political support seems lacking is to suggest that further study is needed. In this instance the matter was referred to a sub-committee of the newly formed Scottish Water Advisory Committee (the Broun Lindsay Committee, hereafter referred to as 'Broun Lindsay'). This inquiry coincided with the implementation of the Rural Water Supplies and Sewerage (Scotland) Act 1944 which, for the first time, brought government subsidies for the provision of improved sewerage and the involvement of engineers employed by central government in the design of schemes of regional drainage.

THE CONTEXT OF A NEW LAW OF POLLUTION CONTROL

While practically all communities of any size were served by public sewers of some sort, there were smaller villages where the only systems had been privately installed and there were still large numbers of homes that lacked water closets. Many sewage treatment works were overloaded and improperly maintained. DHS, in a similar fashion to its work with respect to rural water supplies, had devised schemes of sewers and treatment works, and had taken the recommendations of the Royal Commission on Sewage Disposal, 1915 (CMD 7821) as standard when deciding the nature of treatment required. Generally, sewage treatment works (STW's) offering more than primary treatment were recommended for communities of over 500 people whilst sedimentation alone or septic tanks would suffice for smaller communities. Almost all county councils had schemes under consideration in the late 1940s, but, as with water supplies, a shortage of labour and materials limited the pace of progress well into the 1950s.

Developments in Dumfriesshire and Aberdeenshire were typical. The scattered distribution of settlements in Dumfriesshire limited the scope for regional sewerage schemes. Instead, the County had 40 highly localised improvement plans estimated to cost a total of £193,000. Each served a single village and consisted of a new system of main sewers intercepting existing private drains and leading to small treatment works, ranging from septic tanks to filters. A similar pattern applied in Aberdeenshire where fifteen villages were provided with sewerage and a STW where no facilities had existed before. But Aberdeenshire also had a scope for regional schemes: the outfall from three communities on the Lower Dee, the contents of only one of which received any treatment, was intercepted and led to join the City of Aberdeen's sewerage system which in turn led to a sea outfall without treatment (Burns, 1952; Reid, 1956).

Although engineers from DHS were involved in an advisory capacity in both types of scheme, water supply and sewerage developments at this time were not always co-ordinated. Caithness, for example, benefited from government grants for regional water supply in the 1950s (Baker, 1965). As piped water supplies were improved, and as consumption increased, so too, did the need for proper facilities for waste disposal. The landward area of the county had had virtually no sewers before 1934, but it was not until 1961 that a start was made on new sewerage systems and STWs for the twenty villages in need of them (McIntosh, 1966).

This sequence of priorities appears typical over large parts of Scotland but in some of the urban areas a different view was necessary. The large-scale developments in prospect in the post-war planning boom focused attention on sewerage as a serious problem in places. Systems were generally inadequate to take an increased volume from new housing estates with their improved sanitary arrangements, let alone industrial developments. In the Clyde Valley nineteen of the thirty-two burghs had no reserve capacity and the position was the same in half the Special Drainage Districts (Abercrombie, 1949).

Ayrshire was one of the better counties. Turing, in an independent survey in 1949, pointed out:

> Thirty years ago industrial waste and domestic sewage had converted many (Ayrshire) rivers into something approaching open sewers, but of recent years a great improvement has been seen. The credit for this is due primarily to the energetic action of the Ayrshire County Council officials who have taken up the case of pollution with skill and address and have schemes for the improvement in operation or in hand which will go far towards restoring these beautiful streams something of their former purity. One main valley sewer, on the Irvine, is already in operation and others ... are planned for construction as soon as labour and materials are available.
> (Turing, 1949, p. 32)

Lanarkshire County Council officials were also praised for being "commendably active" but attention was drawn to unnecessary pollution at Hamilton. Turing noted that:

> For some unexplained reason the government has refused to allow the necessary expenditure on a full scale treatment plant and incredible as it may seem, the sludge is actually being passed into the river. The absurdity of such affairs seems hardly to need comment.
> (Ibid., pp. 49-50)

Clearly, resources were not available to improve housing, water supplies, sewerage and sewage treatment all at the same time and the latter lost in the scale of priorities wherever it did not actually represent a bottle-neck to development in other spheres of activity. In this context a strong code of new measures towards the improvement of river water quality could be a positive nuisance.

THE BROUN LINDSAY COMMITTEE ON RIVER POLLUTION PREVENTION IN SCOTLAND

Broun Lindsay published its conclusions in October 1950 (DHS, 1950b). It confirmed most of the recommendations of the ACRPP but was notably silent on a more positive role for local authority sewerage departments in the form of systematic plans for improvements and powers to the River Boards to undertake remedial works directly. More positively, Broun Lindsay agreed that, since pollution moved with the stream, the river basin was the most appropriate unit of administration. Authorities who had made a considerable effort, such as Lanarkshire, complained that they had lost control of stretches of river as other authorities (large burghs) expanded their boundaries to accommodate urban expansion in the form of new housing estates and the like. Efforts to control pollution had proved ineffective, partly because the law was inadequate and partly because the units of its administration (county councils and large burgh councils) were too fragmented. Quite apart from anything else, the importance of salmon fisheries on some rivers warranted 'source-to-sea' control. In evidence it had taken, representatives of angling and fishery interests all pressed the adoption of the ACRPP's suggestion of single-purpose boards for the control of river pollution.

The Rivers (Prevention of Pollution) Act of 1876 was now out of date in the light of developments of sanitation, scientific research and "new ideas of planning and the location of housing and industry" or, indeed, the "claim of the community generally to share in the country's natural heritage" (DHS, 1950b, p. 11). The Committee felt that it was now technically possible to regard pollution from domestic sewerage as inexcusable and to adopt the rule that pollution need be tolerated from industrial sources only if the best practicable means of treatment had been provided. These views should cope with most sources but, further, any untreatable wastes could be accommodated by using the new Town and Country Planning procedures to direct waste-producing industry to places, such as coastal sites, where the receiving waters could cope by virtue of the dilution available.

The Broun Lindsay Committee thought that standards of acceptable quality for effluents were now possible and that the validity of the appropriate range of tests and measures was proven. The important question, however, was whether such standards could be fairly applied. Discussion on this point had revolved around the question of who should set the standards. Anglers and others interested in rivers had argued for the prescription of a common standard for the whole country by central government, presumably because of the conflict of interest that would exist if local authorities had both responsibility for sewerage treatment works and a mojority vote on the setting of standards.

The Committee felt, however, that instead of River Boards applying to vary their local requirements from a set of national standards, the procedure recommended by ACRPP should be reversed: River Boards should derive standards locally, with provision for their review by central government where and when necessary, thus retaining the supervisory role of central government enshrined in the 1876 Act.

The form of the new code of practice having been established, the next major item was the drawing of boundaries around groups of river basins. The areas recommended by Broun Lindsay are depicted in Figure 4.1. It was believed that the problems of pollution in the Highland counties were such that boards need not be formed and that Highland local authorities could administer the new code. Elsewhere, 'River Purification Boards' (RPBs) would be made up of representatives from the existing authorities responsible for the control of pollution and representatives of interests in the rivers such as agriculture, fisheries, industry and the landowners. RPBs should appoint river inspectors who would be suitably qualified in the identification of pollution and the application of the appropriate standards of water quality and have considerable experience in the operation of sewage treatment works (the source of much of the existing pollution). Clearly, the derivation of suitable standards for the areas of RPBs could not be completed until a full knowledge of the hydrological characteristics of the basins was achieved and, in particular, available dilution was assessed. This would take some considerable time to achieve and hence an interim system of control was thought appropriate for the period in which the bye-laws were being formulated. During this period, it was recommended that only NEW discharges should be made subject to the consent of the RPB, the consent being conditional on appropriate standards of quality being achieved or maintained.

The Rivers (Prevention of Pollution) (Scotland) Act 1951

It was one thing to examine the prevailing situation concerning the control of water pollution and produce a set of recommendations designed to deal with it effectively. It was quite another to get the new system adopted. There was, in the context of post-war reconstruction, a great deal of good will but there were also many other competing items on the agenda for action, not least in Parliament. Could new legislation find a place in the government of the day's programme? Beyond this, there was also uncertainty as to how far the proposals would be acceptable amongst the polluters and the Scottish local authorities.

Fortunately, the Broun Lindsay Committee had enthusiastic supporters amongst several Scottish Members of Parliament. Shortly after publication of its report, Mr McKie (Member for Galloway) promoted a 28-clause Private Members Bill in an attempt to have its recommendations adopted. This bill was subsequently taken up by the Government, although it was said of the Rivers (Prevention of Pollution) (Scotland) (Number 2) Bill (which became the Act of 1951) that:

> ... like its partner for England and Wales we might not have had it but for the very narrow balance in the present Parliament, and the necessity of keeping the House of

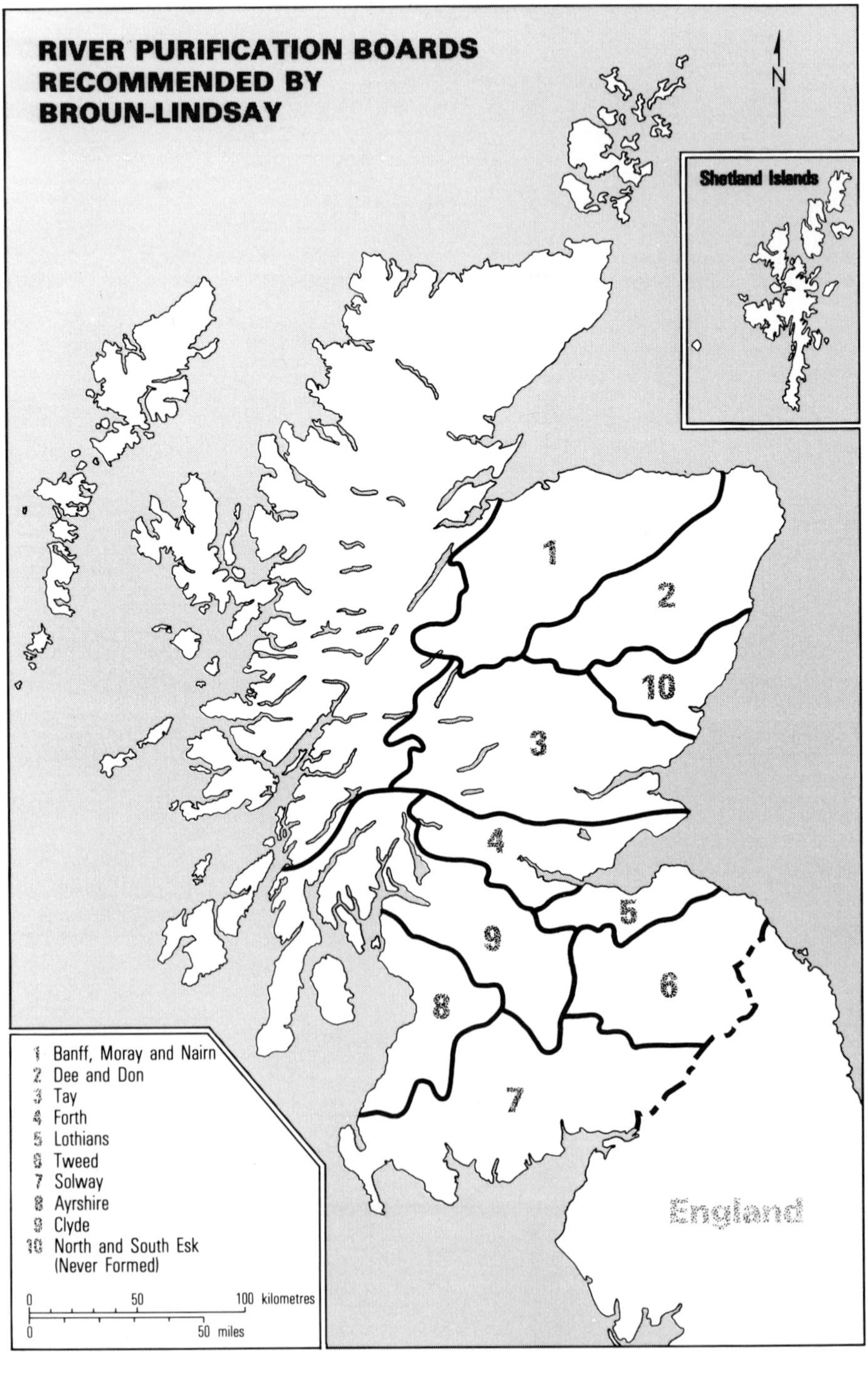
RIVER PURIFICATION BOARDS
RECOMMENDED BY
BROUN-LINDSAY
N
Shetland Islands
1
2
10
3
4
5
9
6
8
7
England
1 Banff, Moray and Nairn
2 Dee and Don
3 Tay
4 Forth
5 Lothians
6 Tweed
7 Solway
8 Ayrshire
9 Clyde
10 North and South Esk (Never Formed)
0
50
100 kilometres
0
50 miles

Figure 4.1

> Commons, so far as is possible, on work of a non-controversial character. But, it is an ill wind that blows nobody any good, and the people of England and Wales and of Scotland are going to benefit, and nobody really minds the reason for their benefitting so long as they do benefit. (Hansard, cols. 2049-2050)

All the associations of local authorities had opposed the creation of River Purification Boards. They did not object to the inclusion of non-local authority interests, they said, but wished to retain the function within their own authorities. In the passage of the Bill through Parliament, however, several MPs took the view that, while they would normally be against the principle of transferring powers from elected authorities to Ad Hoc bodies, this case was an exemption. Prevention of pollution was not purely a matter for local authorities. Industry and others also had vital interests and these could best be accommodated by an Ad Hoc body rather than by co-opting representatives to local authority committees, as some had advocated.

Besides dealing with pollution of inland waters, a majority of the Broun Lindsay Committee had also recommended that all tidal waters should be brought under the control of RPBs Hitherto, no controls had applied to salt water and industrial representatives on the Committee had opposed such an extension. Many local authorities, too, discharged sewage to tidal waters without treatment. The Government proposed a compromise approach: provision was made to include the Clyde and Forth estuaries within the ambit of the new system of controls when the time was right, by Order of the Secretary of State. The latter thought that a Bill that ignored these two estuaries would fail to tackle the very crux of the Scottish pollution problem. However, the necessary capital works could be undertaken only over a period of years and it would, in any event, take a very long time to complete surveys, decide standards and determine where relaxed standards might be allowed, so that the application of the Bill to tidal waters would not be immediate.

In his view: "We have to remind ourselves that our heavy industries, our polluting industries, in the main are on tidal waters for the very simple reason that the sea was the place in which to discharge their effluents" (Hansard, 1951, cols. 1982/83). It would take some time before these industries could deal with their effluents to conform to any standard of the kind envisaged by the bill.

The Rivers (Prevention of Pollution) (Scotland) Act 1951 implemented most of the recommendations of the Broun Lindsay Committee, notably the introduction of a bye-law system for defining offences of pollution, the establishment of an interim consent system, the setting up of River Purification Boards and the employment of river inspectors qualified through membership of the Institution of Sewage Purification. The new code for preventing pollution considerably altered the balance of power among existing interested parties. The angling interest achieved a more formal and influential role through potential membership of RPBs as nominees of the Secretary of State (who would appoint one-third of the members of all such boards to represent wider interests, the remaining two-thirds being nominated by constituent councils previously responsible for the function). District Salmon Fishery Boards might also be represented, but perhaps more significantly for this interest group, the new Act replaced the anti-pollution measures of the Salmon Fisheries (Scotland) Act 1862, in effect transferring to the public purse, the costs of protecting the valuable assets that fishing runs had become. The agricultural and landowning interests became involved for the first time. Industry, although losing some of its formerly extensive rights of appeal, could now, through representatives on the RPB's, advocate a case for reasonableness on grounds where the best practicable measures were already being taken. Furthermore, the procedures for appealing against interim consent conditions and the standards set in bye-laws substantially preserved the principal rights of individual industrialists.

The local authorities, however, lost control. Although authorities had many other matters competing for attention and funds, as indeed did central government, it was now possible for a local body to be forced into action by being outvoted on a matter of effluent standards. The only saving grace was that all such bodies were in the same position, so that members from other authorities could well understand and sympathize with a colleague who felt that his authority's programme of investment could not accommodate expenditure on treatment of sewage at that particular time. Some sympathy for this point of view also seems evident on the part of central government as the ACRPP's proposals for definite plans of improvement had failed to filter through the Broun Lindsay Committee, as had the idea of undertaking schemes of improvement directly in the event of a local authority defaulting.

The most critical aspect of the new legislation, perhaps, was the creation of a corps of statutory river inspectors. For the first time a body of professional men would have the opportunity to make a career out of promoting the maintenance and creation of clean rivers. They would have a day-to-day interest never before seen and have the necessary expertise. Hitherto, the problem of controlling pollution had been tackled on an emergency basis or incident-by-incident by sanitary inspectors who had many other duties to fulfill. Pollution would now become the full-time activity of specialists in that field although progress would always be constrained by the ability of polluters, most notably the local authorities, to afford the scale of improvement that might be necessary, in the face of competing priorities on a crowded agenda.

CLOSING THE CIRCLE OF CONTROL: THE DISCHARGE OF TRADE EFFLUENTS

The successful operation of the RPBs, however, was to be frustrated by a significant loophole in the cycle of control of waste disposal: local sewerage authorities could not begin to plan a programme of improvements if they did not have control of the contents of their sewers and hence precise knowledge of the design performance that might be required of their sewage treatment works. The ACRPP had pointed out as much, but there had been no mention of controls over trade effluents in the 1951 Act.

Broun Lindsay had agreed with the ACRPP that industry should be given the right to discharge to public sewers (as opposed to having to deal with whatever local arrangements were in force in that particular locality). In the context of the new code prevention pollution (which applied to new discharges in the interim period when by-laws were being derived, but in which a significantly altered discharge, whether in volume or quality, would be considered as 'new'), however, this right might well have involved local authorities in a good deal of treatment of effluent that had not previously been necessary. The question was, therefore, whether or not the industrial dischargers should share a part of the additional costs incurred as a result of their new right to discharge whatever they liked to local sewers.

Should industry contribute to the cost of any modifications that were necessary to meet new standards? The issue was regarded as a "separate but related matter" which could be dealt with by later legislation while surveys and the preparatory work towards the by-laws (and the control of non-new or "existing" discharges) went ahead notwithstanding the fact that this question had already been decided in England and Wales in 1937, with no charge being imposed. A critical question concerning the control of water quality was thus once again referred to a committee of enquiry. The Hill Watson Committee (DHS, 1954), consisting mainly of representatives of local authorities and of industry and officially entitled 'The Drainage of Trade Premises Committee', first met in July 1951, but its report was not published until April 1954, its recommendations were not enacted until 1968, and that Act was not implemented in any part of the country until the latter part of the 1970s, once again testifying to the battle of priorities, in which water quality was very much consigned to a lower league in Scotland.

Committee on the Disposal of Trade Effluents

The Hill-Watson Committee concluded that it would be inequitable to expect local authorities to shoulder the responsibility and cost of a duty to receive effluents without qualification. Any such duty would have to be accompanied by the right to demand notice of a trader's intentions, to refuse to admit certain substances, to lay down that reception was conditional on certain requirements being met and to make charges where appropriate.

With respect to charges, the Committee's view was that local authorities would be required to show how the proposed fees related to the expenditure incurred, although this did not mean that a local authority could charge only if works had been carried out and expenditure incurred solely as a result of a trader's application. A charge could equally well be made that related to the cost of works already carried out, so long as account was taken of the proportion of the system that would be occupied by the trade effluent. Local authorities would thus be able to anticipate industrial developments and income from related waste disposal when planning new facilities. It seems fairly clear, however that the Committee did not intend that any system of charges would ever be set at a level that would encourage industry either to recycle its own wastes or to develop processes producing effluents that were less potentially polluting. Indeed, the Committee was well aware that the policies adopted by local authorities were likely to take a quite different form. Local authorities frequently encouraged traders to enter their districts by offering free facilities for the disposal of effluent and by offering free sewerage. The Committee did not wish to see such policies interfered with by national legislation and suggested instead that the statute should be framed so as to leave open the options of providing a free service or levying a charge as seemed appropriate to the local authority concerned.

This question was not, however, to become a real feature of Scottish water management for another thirty years or so, and is discussed again below in chronological sequence. Meanwhile, the attitude of Scottish local authorities to the loss of any function was being clearly displayed during the 1950s through the long time it took to establish the new River Purification Boards, let alone staff them and proceed with suitable research towards the formulation of bye-laws.

IMPLEMENTING THE RIVERS (PREVENTION OF POLLUTION) (SCOTLAND) ACT, 1951

The Broun Lindsay Committee had made concessions to the local authority interest, one of which was the recommendation that the number of seats held by local authorities on each board should be left to the local authorities affected to decide. Yet, having been defeated over the form of controlling authority, some at least, it seems, were not over-anxious to proceed with the formation of new boards. By 1954 only five of the ten boards had been formed, viz., the Tweed, Solway, Ayrshire, Lothians and Banff, Moray and Nairn RPBs. The Forth Board was established in 1955, the

Clyde Board in 1956 and the Dee and Don Board in 1957, while the Tay Board was not formed until 1960. The one remaining board, the North and South Esk Board, was never established. In 1961, the Department of Health announced that attempts to create it had been abandoned and that the County Councils of Angus and Kincardine and the large burgh of Arbroath had been appointed instead as River Purification Authorities for that area (DHS, 1961). Similarly, a long time was to pass before any tidal waters were made subject to pollution controls. The Order of the Secretary of State subjecting new discharges only to controls in the Forth Estuary was not forthcoming until 1960. An Order for the Clyde was published in 1962 but was the subject of objections from local industrialists so that a public inquiry was held into its content in 1963. The report of the latter was made in 1964 but it was not until 1970 that controls were in fact extended to tidal waters of the Clyde, almost twenty years after they had been described as part of the very crux of Scottish water pollution problems.

SECOND THOUGHTS ON BYE-LAWS

Some River Purification Boards had not begun their work at all and most had barely been established when a sub-committee of the Central Advisory Water Committee in England and Wales was commissioned in October 1956 to examine the whole question of industrial effluents once again (CAWC, 1960). An important aspect of this question was the bye-law standards for application by River Boards, the title of river pollution prevention authorities in England and Wales until 1963.

The making of the bye-laws had proved difficult. The English boards had been at work on the problem since 1948 but none had been produced anywhere. Nowhere was there any control over "existing" discharges, i.e., those being made prior to or during 1951. This was a serious matter for the use of rivers as sources and carriers of water supply in England and Wales. On the other hand, the sub-committee had been informed that it had not proved difficult to devise effective conditional consents for new or altered discharges.

"Existing" discharges, which might have been sources of pollution for a very long time, presented very different problems. The sub-committee felt that:

> ... a heritage of years of neglect can seldom be overcome except by patient work for improvement which inevitably requires time, and no useful purpose whould be served by River Boards imposing in respect of any existing discharge such immediate and onerous conditions that those responsible were quite unable to comply with them. Normally it could not even be contemplated to refuse consent outright in such cases.
> (CAWC, 1960, p. 18).

The solution seemed to be that a procedure similar to the consent for new discharges should be applies to old ones with the River Boards having power to attach conditions to consents and to review, and if necessary, vary such conditior from time to time so as to achieve the maximum practicable rate of improvement. The back-log of improvements required, particularly with respect to local sewage treatment works, meant that there could be no overnight introduction of any new code of water quality. Instead, an incremental approach, a one step at a time, point-by-point system of improvement seemed more appropriate. The sub-committee accordingly recommended that provisions for the making of bye-laws in the 1951 (English) Act be repealed and new legislation brought forward to extend the consent system to all sources of pollution. This was done almost immediately in England and Wales where pressure on water supplies was becoming acute in places. The Rivers (Prevention of Pollution) Act 1961 applied to England and Wales only but could they and Scotland have two different codes for the control of water pollution?

The Scottish River Purification Advisory Committee, established under the 1951 Act to advise the Secretary of State on such matters, argued for a similar extension of the consent system to all discharges in Scotland as soon as possible. The Institute of Sewage Purification, whose members included the river inspectors as well as the few managers of sewage treatment works who were fully qualified, issued a memorandum on the urgent need for such additional legislation in Scotland (Institute of Sewage Purification, 1964). Pressure was not acute on rivers as sources of water supply, although the Loch Lomond scheme of water supply was cited as one of the developments that lay behind this call. Instead, in Scotland, more emphasis was placed on an increased awareness of amenities and facilities for leisure, mounting abstraction for spray irrigation and the possibility of increased abstraction of river water by industry.

Another measure of the place of water pollution in the ladder of priorities confronting central government is the fact that the Scottish equivalent of the English 1961 Act did not come before Parliament until 1964 and did not become law until 1965.

The Rivers (Prevention of Pollution) (Scotland) Act 1965

In the debate on the 1964 bill there was little dispute about the need to change the system; MPs seemed happy to follow legislative developments in England and Wales, although the topic of discussion differed markedly from that in 1951. Gone were the traces of optimism of the early post-war period. Instead, the opportunity was taken to bemoan the lack of apparent progress and criticise the performance of preceding governments on the matter.

Mr Galbraith (Glasgow, Hillhead) had been prominent in supporting the 1951 Act but was now disappointed. He thought clean rivers and pollution controls involved "our whole sense of values" but because of that, "quite frankly, I do not expect a very great deal to come out of this Bill" (Hansard, Scottish Grand Committee, 24/11/64, cols. 34-45). He felt RPBs had been afraid to withold consents from big (new) developments because employment in their areas might suffer and in this light were unlikely to rigorously apply new powers over existing discharges. Sir Myer Galpern (Glasgow, Shettleston) was despondent over the contribution of local authorities to improvements, pointing out that the expenditure required was often beyond the means of individual authorities and that central government would have to step in if any significant progress was to be made (Hansard, Scottish Grand Committee, 24/11/64, col. 40).

He saw the standard of sewage treatment by local authorities as the greatest factor in preventing river pollution. Many disposal works were far too small and in some cases were dealing with as much as two or three times their designed capacity. He thought some local authorities were anxious to effect improvements but the Government had not allowed the expenditure to the required levels. "This parsimonious attitude on the part of the Ministry towards sewage disposal is probably the greatest single factor today contributing to continued pollution" (Hansard, Scottish Grand Committee, 24/11/64, cols. 48-49).

Summing up the debate, Dr J Dickson Mabon, the Under Secretary of State with responsibility for such matters, agreed that progress had been disappointing. He saw three reasons for this. First, local authorities had not made the contribution that was necessary, and the Government was considering representations from the SRPAC on how this could be improved. Secondly, there had been the leisurely pace at which the RPBs had been established. Thirdly, there had been the failure to implement the conclusions of the Hill-Watson Committee. It was all very well, however, to proclaim the desirability of clean and pollution-free rivers but "one must realise that there must be a practical balance" (Hansard, Scottish Grand Committee, 24/11/64, col. 87). Industry was essential to Scotland and it would be wrong to place obstacles in the way of new industry or to hamper the prosperity of existing industries. A balance, then, had to be struck between the needs of industry and the public's need for higher amenity. Local authorities also had to balance the claims of sewage treatment with those of housing and other social programmes.

Hence, the River Purification Boards began to try to effect some control over the quality of river waters in 1966, well aware that a slow improvement was expected of them. The extent to which radical changes were necessary in the management of sewerage and sewage treatment permitted no other option.

DIFFICULTIES WITH LOCAL AUTHORITY SEWERAGE AND SEWAGE TREATME

In an assessment of the position in 1968, the Institute of Water Pollution Control (the Institute of Sewage Purification retitled) reported the nature of deficiencies to a Royal Commission on Local Government (the work of which is extensively considered in Chapter 5) (Wheatley Written Evidence, 1967a). There were too few regional sewerage schemes because of local jealousies and "irrelevant and trivial" disputes between authorities. Difficulties had arisen because of:

1. ineffectual agreements over the reception of trade effluents at sewage treatment works, so that the works could not perform properly;

2. competition between local authorities for new industrial employment, resulting in factories in unsuitable locations with, subsequently, difficult and costly problems of waste disposal; and

3. trade effluents that could have been accepted without difficulty at larger or centralised plants being fed to small under-equipped works.

Many sewage treatment works suffered from unsatisfactory design, unsatisfactory management, inadequate capacity or inadequate operational supervision. In short, the sewerage and sewage treatment services required a complete overhaul in terms of their management structure and their physical facilities.

The fragmented structure of local authorities had had two serious effects: firstly, a weak financial base in many authorities badly affected the design and implementation of new schemes, maintenance and technical supervision; and secondly, there was a serious lack of specialist staff.

Equally grave was the continuing lack of a systematic approach to the problems of receiving trade effluent at sewage treatment works. The administrative structure of local sewerage authorities came under review in the late 1960s as part of a move towards local government reform as a whole, which is considered in the following chapter. Immediate action on trade effluents was forthcoming in 1966.

The Labour Government (1966-1970) had intended to take up the question of trade effluents in that year but the urgent and apparently unexpected need to legislate for regional schemes of water supply (through the Water (Scotland) Act, 1967 considered in Chapter 3) delayed these plans until 1968.

The Sewerage (Scotland) Act 1968

In introducing the Bill to Parliament in January 1968, the Secretary of State presented its measures as "really a further step in our measures for modernising Scotland" (Hansard, Scotland Grand Committee, 23/1/68, cols. 5-12). Although the nineteenth century Acts and the jumble of local legislation that was now to be replaced had been useful, there were still several irritating anomalies and uncertainties. For example, there was no certainty over who should provide sewers for new housing estates (the sewerage authority or the builder) and practice varied from area to area. A modern and more systematic code of legislation was needed to help local authorities plan ahead for the provision of sewerage to serve both public and private housing estates and industrial development. Henceforth all local authorities would have a statutory duty to provide the sewers. This certainty would help planning but at a price and, because of the cost and in light of the then-current economic situation, the Secretary of State did not propose to implement the Act and its new duties until such time as an increase in public expenditure could be afforded.

The Act is in two parts. The first provided a systematic code of rights and duties concerning the provision and maintenance of sewers. The second implemented the Hill-Watson Committee's recommendations concerning a standard code of practice for the acceptance of trade effluents into local authority sewers and sewage treatment works, subject to appropriate conditions being set by the local authorities and the possibility of levying charges for the reception of effluents.

Part Two, at least in theory, could have been self-financing, but Part One could not and, because restrictions on capital expenditure remained in force, the Act was not implemented until 1973. By this time the local authorities were on the point of being reorganised and seem to have postponed any innovative action because responsibility was about to pass to new regional councils. These in turn seem to have had other problems of a higher priority to deal with in their first few years so that it was not until 1977 that Part Two was fully implemented. Lothian Regional Council's scheme of charging for trade effluent is considered as part of the discussion of expanding the range of choice of water management strategies in Chapter Six.

AN OVERVIEW OF THE BROADENING APPROACH TO WATER QUALITY MANAGEMENT

As with water supplies, it is clear that the need for action was recognised very early on. In 1936 the Advisory Committee on River Pollution Prevention had elucidated the administrative problems and had outlined what was necessary. Over the following forty years these proposals were implemented in a disjointed and incremental manner.

The concept of river basin management was the first to be adopted but the doubts of local authorities over its utility meant that one River Purification Board was never formed and the remaining system of nine was not complete until 1960. The title "River Purification Board" is a misnomer because the original recommendations to confer on them direct powers of improvement were never carried forward, probably because of hostility by local authorities to the ideas of losing control over capital expenditure, and the boards' intended roles as co-ordinators of purification plans were not forthcoming. Behind this appears to be a low priority for expenditure on sewage treatment, the provision of sewerage taking by far the larger part of the joint budget for sewerage and sewage treatment until the late 1960s Both local and central governments appear to have taken the view that the improvement of servicing housing and development had to take priority over water quality. The prevailing attitude appears to have been that, while people are unlikely to forego a supply of wholesome water or decent housing without protest, they may well tolerate delays in the provision of facilities for the treatment of sewage or polluted rivers for a few years more.

In this context a full range of controls over sources of pollution did not come into force until 1966 and a beginning was not made on the radical reorganisation of sewage treatment until 1975, after local government reform. Further no effective control over "existing" discharges to tidal waters came until the 1970s and then as a result of the influence of contemporary changes in the law of pollution in England and Wales and within the European Economic Community. These are also considered in Chapter Six.

The shadow of contemporary developments in England and Wales has been a consistent theme in this chapter. There are no instances of Scottish practice influencing in a major way colleagues south of the Border. We may conclude that, unlike England and Wales where increasing pressure on water resources for purposes of water supply presented new problems and stimulated a broadening of the approach to water quality management, that broadening came in Scotland as a result of the English dimension in Scottish legislation and may be seen as eccentric to the main themes of water management in Scotland, the subject of the next chapter.

5

Scottish water management reconsidered

An opportunity to review the structure of water management in Scotland was provided in 1966 by the Royal Commission on Local Government in Scotland, chaired by the Rt. Hon. Lord Wheatley and appointed just four months after SWAC had published its final proposals for regional water boards and a year before the enabling Water (Scotland) Act passed into law. Water supply, sewerage and sewage treatment were not, of course, the main concerns of the Commission, but they were functions of local government, and, as such, were to be considered in a broader perspective of policy planning, structure planning and strategic planning, along with transport, housing and education. The Commission's inquiry, which (like all such inquiries) provided published statements of views by a wide variety of bodies, public and professional, was also an opportunity to reconsider the place of river purification generally, which had been removed from local government in 1951 with the formation of the River Purification Boards.

At the root of the movement to reform local government lay the fact that the geography of Scotland had changed dramatically over the forty years since the structure of local government had last been examined in 1929. Administrative boundaries were now out of date in the light of suburban development, often dividing centres of population from hinterlands with which they had close functional connections, and sometimes even splitting continuously built-up areas. As with water supply, there was an increasing feeling that many authorities were too small, with insufficient staff and financial resources to manage all their functions to the greatest effect. Local services were sometimes substandard and over the years, a series of *ad hoc* or joint administrative arrangements had grown incrementally as particular problems had been tackled on an individual basis. Having intervened to reap the benefits of a broader view, the Scottish Office was felt by some to have gained a *de facto* level of control over local affairs to the detriment of local politicians and local control over those affairs generally.

The Wheatley Commission (hereafter referred to as 'Wheatley') had the basic difficulty of resolving, on the one hand, demands for larger operational areas for the management of services in the name of efficiency and economies of scale, and on the other, demands for a return to a system of local authorities that would be lively and strong foci of local political interest. Individual functions and services were not the Commission's primary focus of interest which was rather the system as a whole. Within that remit, the needs of 'planning' of different kinds seemed paramount. Three forms of planning are identifiable: activities undertaken by virtue of the Town and Country Planning Acts, such as the location of industrial estates and the direction of urban expansion; planning in its broadest sense of policy-making, including finance and corporate planning; and 'strategic planning' which was taken to mean the planned use of resources and the direction of public investment in accordance with a strategy of national and regional development (Wheatley, 1969, pp. 58-59).

The spirit of the time was very much in favour of the last so that Wheatley could take the view that the ordinary Scot's greatest concern was with his economic environment and hence any attempt to revitalise local government should pay close attention to the so called 'implementing services'. These were the components of strategic planning: the construction of 'advanced' factories; the provision of roads, public transport, utility services, including water and sewerage, as well as the promotion of schemes of urban development and renewal or the designation and development of new towns. These views were very much in line with the then-Labour government's policy, with its creation of a new Department of Economic Affairs and the production of Great Britain's first (and last, to date) national plan (CMND 2764, 1965). They were not, however, compatible with the essentially nineteenth century structure of local government that existed, which in turn meant that many key activities were approached in a partial, fragmented and unsatisfactory manner. Any idea of partnership between local and central government was thus not realised, with the latter playing by far the dominant role.

To derive a general schema, some knowledge of each function was obviously necessary but the detailed characteristics and problems of each function were not Wheatley's concern. Written and oral evidence was therefore taken on which an overall view could be based. The Commission could then concentrate on the inter-relationships between functions and the question of the most appropriate administrative areas. The evidence given to them nevertheless serves the present purpose admirably by giving a contemporary view of the experts' opinions about what was most appropriate for the various water-related pursuits.

WATER SUPPLIES

The Wheatley Commission took evidence on the water supply service on two occasions. Written evidence was submitted by the Scottish Development Department (SDD) and the Institution of Water Engineers before SWAC's scheme for 13 regional water boards had been enacted by virtue of the Water (Scotland) Act 1967, whereas oral evidence was taken from both, AFTER the Act had passed through Parliament (Wheatley, Written Evidence, Vol. 7, 1968. In this context, the evidence offered was inevitably somewhat contradictory and confused, as civil servants could not comment on matters about to go through Parliament and the Institute had little choice but to accept the newly legislated form of operations for the service after it had received the Royal Assent.

Focusing on the scene before 1967, SDD explained that SWAC had been considering a reorganisation of the administration of water supplies for a number of years and that it had recommended a system of about a dozen regional authorities, the boundaries of which, generally speaking, had been drawn with regard to topographic features, so as to include within their areas the main sources of supply which they might need to develop in future. Meanwhile, SDD officers (though not specifically including the Chief Water Engineer) also argued for the adoption of a regional perspective for both planning and implementation of infrastructural services, such as housing, water, sewerage and roads. The regional scale was necessary for all such activities because the development of major projects had been hampered on several occasions by disputes as to why individual local authorities in the general area of a major new development should be required to pay for the housing or the water necessary for that project, whilst the revenue in terms of local taxes and the boost to the local economy provided by the project went either to another specific local authority or to a group of them. (Wheatley, Written Evidence, Vol. 7, 1968).

At first sight the two sets of evidence from SDD appeared to be contradictory: when considering planning and regional development, water was seen as an implementing service and as such suitable, indeed necessary, for regional units of administration to replace the existing patchwork of local authorities; when viewed specifically from the perspective of securing the future water supply of various parts of the country a regional framework was also deemed necessary. But were the two regional frameworks of the same sort, a question the Commission asked when oral evidence was given by the Department. The answer given by the Secretary of SDD, Mr. A.B. Hume, speaking in support of SWAC's recommended regional boards, was that water catchment and general local authority boundaries were incompatible. He pointed out that SDD water engineers had compared the needs of water supply with the eight regions proposed for planning and infrastructure services by the Regional Development Division and had found 'serious conflicts of boundary.' (Wheatley Oral Evidence, 20/6/67, QQ 4090-4092).

The Institution of Water Engineers (IWE) was not prepared to accept that SWAC had recommended the optimum solution to problems of water supply. It argued that "previous remits (to SWAC) have been too resticted and too narrow and we have tried to look at the whole of Scotland now ... we must clear our minds of the existing proposed boards (SWAC's 13) for which draft orders have been prepared and/or published". When its witnesses gave evidence on Central Scotland they proposed one board which could be divided into areas for administrative purposes: "We consider this to be the correct solution for the whole country but, if that is not acceptable ... much could be achieved by the following proposals ... six boards would be sufficient ... their boundaries ... drawn from a relief map on topographical considerations ..." (Wheatley Written Evidence, Vol. 11, 1968, pp. 56-59). The boundaries of the areas to be covered by such boards are depicted in Figure 5.1.

The IWE wanted water taken out of local government altogether. Its favourite proposal was to have water made the subject of single-purpose management across the country. Local government, it felt, had too many other interests. When a conflict of interest arose, experience had shown that the different committees of a local authority would reach a compromise not necessarily in the best interests of sound water management. When representatives gave oral evidence in February 1968, however, SWAC's system of thirteen regional water boards was being implemented. The IWE then had no choice but to reiterate its belief in a single Scottish water authority as the best long-term solution, but accepted that SWAC's thirteen regional boards were satisfactory at the present stage. The fact that regional boards would not be under the direct control of local councils would, in the view of witnesses, ease the problems of improving the water service, the witnesses thought because that required the provision of large amounts of capital and, hitherto at least, very few local authorities had been prepared to do this.

The Commission asked why, if a new structure would have much larger, financially stronger local authorities, the service could not be returned to them. The witness responded that the latter would have so many other things to do that they would not look on the service in a single-purpose manner and that it would inevitably suffer (Wheatley Oral Evidence, QQ6468-6559).

EVIDENCE ON POLLUTION CONTROL AND WASTE DISPOSAL

As noted in Chapter 3, SDD had published a White Paper on the modernisation of local government in 1963. In that paper there had been no specific discussion of water management, but water supplies and river purification were both included in the list of functions thought best performed by the then suggested 'top tier' regional authorities. The

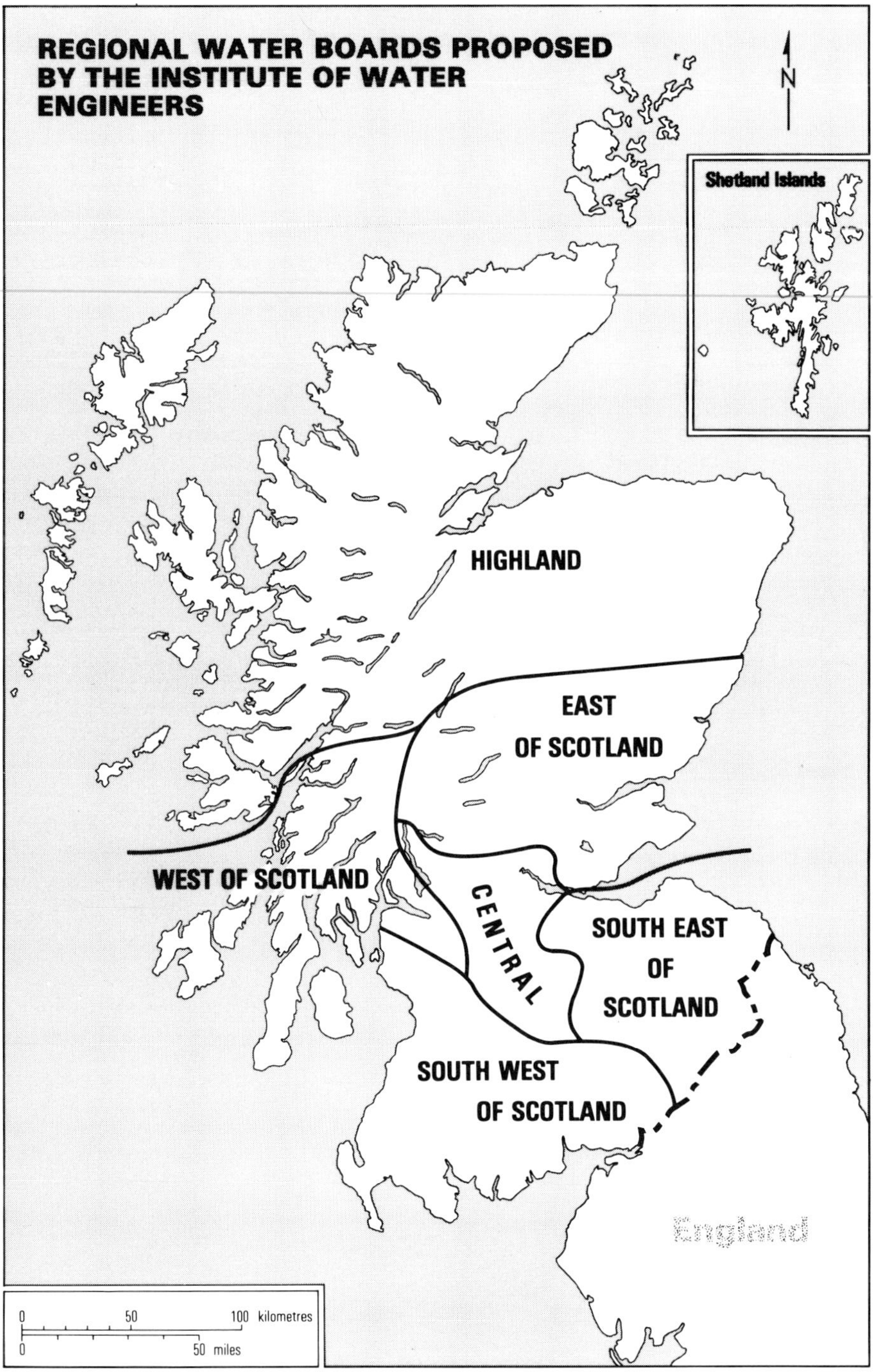

REGIONAL WATER BOARDS PROPOSED BY THE INSTITUTE OF WATER ENGINEERS
N
Shetland Islands
HIGHLAND
EAST OF SCOTLAND
WEST OF SCOTLAND
CENTRAL
SOUTH EAST OF SCOTLAND
SOUTH WEST OF SCOTLAND
England
0
50
100 kilometres
0
50 miles

Figure 5.1

various witnesses who appeared before the Wheatley Commission, however, tended to differ from this view, particularly with respect to the prevention of pollution.

The Scottish Development Department claimed, for example, that the existing organisation for the prevention of pollution seemed generally well suited to its task, with the boundaries of the River Purification Boards (RPBs) having been determined by topographic features to include the watersheds of the main rivers, a situation which was unlikely to suit any of the other functions of local government.

SDD also suggested that valuable economies of scale would follow if sewerage and sewage treatment were administered by much larger authorities than hitherto (Wheatley Written Evidence, Vol. 7, 1968, p. 21; Wheatley Oral Evidence Q 4679). While drainage generally followed the lie of the land and river catchments might therefore be thought of as the 'natural' areas to select for the purpose of pollution control, such areas were not essential for the management of water supply and sewerage. Moreover, staff concerned with these latter services were interchangeable to some extent. There would, therefore, be some advantage in allocating sewage management to the same agency as water supply, especially outside the populous Central Belt. Facilities could be shared and this would to some degree alleviate the generally-recognised problem of the unsatisfactory management of smaller works because of the lack of skilled supervision. SDD also argued that the RPB was a prevention authority for the river basin and, while ideally a sewerage authority for the same river basin might undertake the practical work of purifying sewage before it came into the river, that was a different type of function, requiring the same sort of staff as water supply. SDD witnesses therefore thought that a combination of water and sewerage functions might be a more practical solution than a combination of RPB's and sewerage authorities.

The Commission could not understand why SDD thought that sewerage could more appropriately be associated with water supply than with river purification, for river purification and sewage treatment were surely parts of the same problem. SDD witnesses replied that RPBs were enforcement authorities whilst sewerage authorities were practical treatment agencies. Sewerage authorities were not always efficient and that was one reason why RPBs were necessary. They also pointed out that RPBs included representatives not only from local government but also from interests whose livelihoods were affected by what the RPB might require. This factor would make it difficult to include the work of RPB's within any new structure of local government if the hitherto-accepted approach to pollution control, namely negotiation, were to be continued (Wheatley Oral Evidence, Q 4862).

The Institution of Water Pollution Control took the view that the physical pattern of future trunk sewers and sewage treatment works should be determined by topography rather than by local authority boundaries. The operations of autonomous sewage authorities had given rise to schemes which, in retrospect, appeared unduly costly and restricted. Furthermore, the law of the land was "frequently being disregarded both in action and spirit". and not only had the rate of improvement in the quality of many rivers been disappointingly slow but in places actual deterioration had occurred. Like the Institution of Water Engineers, this Institution was not impressed by the record of existing local authorities (Wheatley, Written Evidence, Vol. II, 1968). With the support of the Confederation of British Industry in Scotland, it took the view that the best way of spreading the costs of treatment, while at the same time obtaining the best results, would be through the creation of regional drainage authorities. The Association of River Inspectors also favoured such a move, reminding Wheatley that one of the reasons why RPBs had been created in the first place was that previous legislation had been administered quite ineffectively by local authorities.

Thus the Wheatley Commission heard evidence suggesting that the function of water management required special, *ad hoc* units of administration, albeit in a variety of different arrangements. But, *ad hoc* bodies were something that the Commission was determined to avoid in their restructuring of local administration in Scotland, a view for which some considerable support had come in evidence from the professional bodies of local administrators.

EVIDENCE ON THE EFFICACY OF JOINT ADMINISTRATIVE ARRANGEMENTS

The Society of Clerks and Treasurers in Scotland saw the main problems of joint committees as those of power without responsibility, and of one constituent member being a dominant party and, as such, taking charge. In its hearings, Wheatley put the following statement to representatives of the Society:

> It is said that the principal objections to joint committees were, firstly, that the nominated representatives from the various local authorities do not really go as full representatives of the body, but merely as delegates of their organisation; they are therefore not really in the full capacity of elected representatives dealing with the overall problems of the particular function. Secondly, it is said that the system of requisitioning leads to a certain amount of irresponsibility in the sense that they are not immediately responsible to the electorate for the amount of money that they requisition from the constituent authorities. We have found in various places through local government, objections to joint committees on these two grounds. (Wheatley Oral Evidence, Q 1044)

The practice of requisitioning, where people had the responsibility for spending money without direct responsibility to the electorate, was severely criticised by the Society. Representatives from burghs, on committees of county councils responsible for some services in the burghs, were believed to have acted irresponsibly on occasion, since they did not have to defend their decisions in the light of the ultimate level of county council rates. At the same time burgh councils resented the requisitions made by counties on them to pay for such overlapping services, particularly education. Not surprisingly, therefore, the Convention of Royal Burghs, in particular, thought the practice highly unsatisfactory. The Association of County Councils thought that it would be advantageous to have units of local government responsible for the whole range of functions appropriate to that scale rather than have multifarious boards operating over different areas (Wheatley Oral Evidence, Q 256).

In the light of such arguments the Association of County Treasurers felt that the water boards should be fitted into the pattern of top-tier authorities. In its view, the 1967 Act had put the cart before the horse and the Convention of Royal Burghs agreed that if technical difficulties could be overcome, it would be desirable to bring water within the same unit of local government as the other major services (Wheatley Oral Evidence, Q 1676). Thus, in the main, the opinion of officials and councillors from existing local authorities was that the provision of water services should be included in the responsibilities of the new units of local government. This was largely because of the general feeling that requisitioning and joint committees were twin evils which should not survive the transition to a new system.

Wheatley agreed and proposed a pattern of six regions for the implementing (infra-structure and planning) and protective (police and fire) services (Fig. 5.2). For the remaining functions, 37 districts were proposed, 10 of which had more inhabitants than the smallest region, the South-West, showing, according to Honey, that Wheatley was "much more interested in matching the spatial organisation of Scottish life than in meeting a functionally defined population threshold." (Honey, 1977, p. 113).

THE CONCLUSIONS OF THE WHEATLEY COMMISSION RELATING TO WATER MANAGEMENT

Without the benefit of professional backing for its views, the Commission proposed that water, sewerage, river purification and flood control should all be direct functions of local government (Wheatley Oral Evidence, 13/3/67).

The Commission was aware that this was "a more robust line" than most of the witnesses had taken, but saw great advantages to be gained through planning the development of these services in combination with one another and with

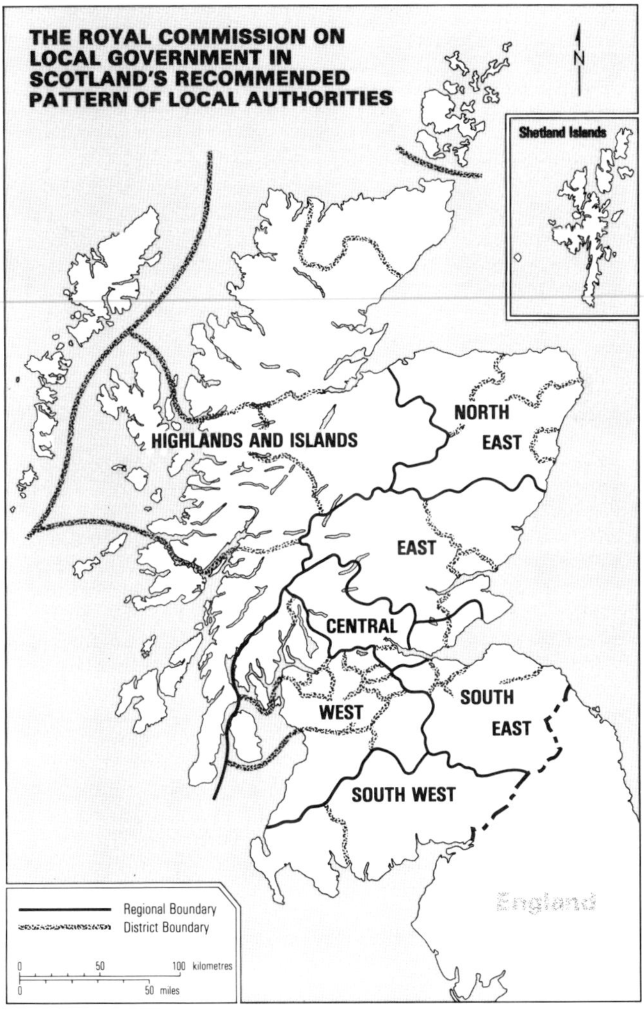

Figure 5.2

other local government services. Benefits, it was claimed, would also follow the bringing together of the technical skills which the services required.

The Commission was also aware of practical difficulties that might arise. In a similar way to the 'added area' device for the administration of water supplies, where if technical considerations demanded it, a neighbouring water authority managed part of another water authority's area of supply, administrative arrangements would have to be devised so that one local authority could look after another local authority's part of a river basin for purposes of controlling river pollution. But the extent of such problems would depend on the pattern of authorities finally implemented. As a Royal Commission it was not at all certain that recommendations would be acted upon and, in this light, it should be noted that the Association of County Treasurers' criticism of the regional water boards, as representing the cart being put before the horse, was slightly unfair, in that there was no guarantee that the reform of the whole of local government would come about in less than ten years whilst the problems of water supply that the boards had been devised to tackle were relatively urgent, as Chapter Three has shown. To the surprise of some, however, the basic framework recommended by the Wheatley Commission was enacted, with amendments, in September 1973, within five years of its report being published.

Notwithstanding the fact that the fruits of its previous labours had had a mere two years to show what they could do, SWAC was re-constituted in January 1970 to consider the Wheatley Commission's proposals. Some urgency was brought to this, the third review of the institutional structure of water supply management in seven years, by the election of a new (Conservative) government in June 1970. The new Secretary of State for Scotland followed his English counterpart in setting the modernisation of local government as a high priority and within nine months published a White Paper on "Reform of Local Government in Scotland" (CMND 4583, 1971). In this document the clear aim was to introduce a Bill in the 1972-73 Parliamentary Session so that councils could be elected in 1974 and the new system become operational by 1975. A similar timetable was to apply to a different set of proposals for England and Wales although these countries would be dealt with first, one year earlier at each stage. The new government took the view that all the materials for a decision were now at hand and that the time had come for action to put an end to the uncertainty which had been in the air since 1963. The content of the 1971 White Paper was viewed as a prescription for action, not a basis for negotiation. Hence it was quickly followed by the Local Government Bill of 1973 containing a clause disbanding the River Purification Boards and transferring their functions to new regional councils.

CONTROVERSY OVER THE STRUCTURE OF SCOTTISH WATER MANAGEMENT

At the very time this proposal was published, the Royal Commission on Environmental Pollution (RCEP) was examining the problem of water quality in a number of British estuaries, including the Forth and Clyde, and took some evidence on it at hearings held locally (Royal Commission on Environmental Pollution, 1972. In 1972, RCEP published its view that the inclusion of the prevention of river pollution within the remit of new local authorities in Scotland would be a retrograde step.

While RCEP would have preferred to see independent authorities responsible for the whole water cycle, as was then proposed for England and Wales (since it was impossible to match the proposed boundaries of a new structure of local government there with those required for water resource planning and development), it suggested that the advantages of retaining River Purification Boards in Scotland as semi-independent *ad hoc* agencies would outweigh the advantages of integrated control under the aegis of local government, particularly since less emphasis was required in Scotland on the links between the control of sewage disposal and of water pollution to enable additional water resources to be developed.

RCEP had heard the Confederation of British Industry argue strongly the case for retaining RPBs, at first sight an unlikely source of support for the river inspectors, who had provided the main body of evidence. While the Confederation understood how the unification of sewage disposal and river purification under the new regional councils might be justified on grounds of administrative tidiness, local authorities had been consistently among the worst offenders and there was a danger that, if they were again given responsibility for the control of pollution, there would be a considerable slackening in the rigour of regulation, even if the local authorities were the subject of increased statutory control by central government. The Confederation also thought it would be unfortunate if the authorities responsible for controlling river purification were to lose the valuable contribution at present made by the Secretary of State's nominees, including their own members, as independent representatives of agricultural, fishing and industrial interests.

Thus, two separate Royal Commissions offered the Government conflicting advice on the matter of pollution control in Scotland. The more general, the Wheatley Commission, had reported first, but within a year of its recommendations being adopted as policy these were being criticised by the specialist Royal Commission on Environmental Pollution. In this context Government spokesmen found themselves in an invidious position, but apparently undaunted, J.W. Shiell, SDD's Chief (Water) Engineer, reviewed the Government's plans for the RPBs in a paper given to the Institution of Water Pollution Control's Annual Conference in 1972 (Shiell, 1973, pp. 261-272).

Mr. Shiell was aware that the Government's proposals had their critics but, after listening to all the arguments, he was not persuaded that there was any sound basis for many of the fears that had been expressed. In principle, the proposals in Scotland differed little from those that had received considerable acclaim in England. All functions relating to water, namely conservation, supply, distribution, sewerage, sewage treatment and water pollution control, would, he thought, be performed by a single authority in each of the countries. There was no reason to suppose that, in the different needs and circumstances prevailing, the system in one country would be any less effective than that in the other. Neither was there any reason to suppose that adequate funds would not be available to enable the authorities on both sides of the Border to perform their function satisfactorily.

Inevitably, there was opposition to these views, particularly from those involved in the existing institutional arrangements. Mr. George Sharp, Chairman of the Scottish River Purification Advisory Committee, for example, was convinced that the decision to place both river purification and sewage disposal under the new regional councils would in fact put the clock back in Scotland (Sharp, 1973). He expanded his views in an article in *Municipal Engineering* entitled 'Scotland's Water Problems Treated as Second Best'. The proposals of the Wheatley Commission were not, in his view, unified sets of proposals, designed with the best management of the water cycle in mind. The non-coincidence of boundaries would often make it impossible for a regional council to proceed independently with river purification measures without having regard to its neighbouring region or regions.

Beyond this, he observed, there was no proposal to unify aspects of the water cycle by making special financial arrangements. He noted that the proposals were being opposed by both the RPB's and the regional water boards, which would be swept into limbo in spite of the good work they had done and the reasons for their formation in the first place (outside the structure of local government) being patently obvious. Mr. Sharp noted that the Government had said that the new regional councils would have greater technical and financial resources. No one was going to dispute that, but many people would doubt the willingness of these authorities to act urgently and spend on services, such as those relating to water, which were never seen by electors and candidates as 'vote catchers'.

PRESSURE TO REVERSE POLICY ON POLLUTION PREVENTION

In the event, it was probably a successful campaign originating among the professional officers that prevented the full adoption of Wheatley's proposals for water management. Four lines of argument were stressed in an attempt to reverse government policy after the proposals in the White Paper of 1971 appeared unchanged in the Local Government Bill of 1973 in spite of the conflicting advice of the RCEP. First, there was past experience of the difficulty of getting local authorities to take remedial action. There was wide support for the view that the RPB's could assist local authority departments of sewerage to maintain a fair share of spending by retaining their status as external and independent bodies acting as pressure groups.

Secondly, the local authority sewerage departments, with which the river purification service would presumably be linked, were generally thought to be responsible for the major part of the pollution that the river purification boards were trying to prevent. It was argued that one section of a local council was hardly likely to prosecute another section or even department of the same council, and fears were expressed over the possibilities of 'cover ups'.

Thirdly, it was agreed that river purification must be managed over the whole of river basins if problems of co-ordination between authorities were to be avoided, and difficulty was expected in matching the boundaries of river catchments with those of the new regional councils. This was a problem, however, which appeared less important when it was examined closely. Non-contiguous boundaries were generally to be found in the unpolluted upland. There was, however, a major problem over the Forth Basin, particularly its estuary, and this was aggravated by the subsequent amendment to the Local Government Bill to create a Fife Region (see Fig. 5.3). Further, the system of shared control ran counter to a simultaneous recommendation from the Royal Commission on Environmental Pollution, that major estuaries should be the subject of unified control, and made the adoption of special arrangements of a joint committee type for East-Central Scotland almost inevitable. If the principle laid down by the Wheatley Commission had to be breached in this, the second most important area of Scotland, it was reasonable to ask why it should be preserved elsewhere.

Finally, many professionals were concerned that the highly-regarded and 'useful' nominees to the RPBs made by the Secretary of State to represent other interests, might be lost altogether. The Local Government (Scotland) Bill empowered the regional councils to make their own arrangements for the representation of affected or 'interested' parties. It may be that the lack of specific provisions in the Bill to retain this element in the system caused unnecessary confusion and provided an opportunity for some opponents to suggest that there might be no representation of industrial or other interests. This confusion was understandable, if unfortunate, since the Bill was, after all, primarily concerned with something else and draughtsmen presumably envisaged that the Scottish River Purification Advisory Committee would subsequently be asked to consider what specific arrangements should be made.

Many of these arguments had featured at the time of the original formation of River Purification Boards but these had now a decade or more of operational practice behind them and the perceived success of this gave them a new vehemence and a new set of vociferous champions, the river inspectors and sympathetic board members and chairmen.

THE LOCAL GOVERNMENT BILL IN PARLIAMENT

When the appropriate clause of the Local Government Bill came to the committee stage, with matters of principle having been discussed, members of Parliament had the opportunity to go through the Bill line by line, and an amendment

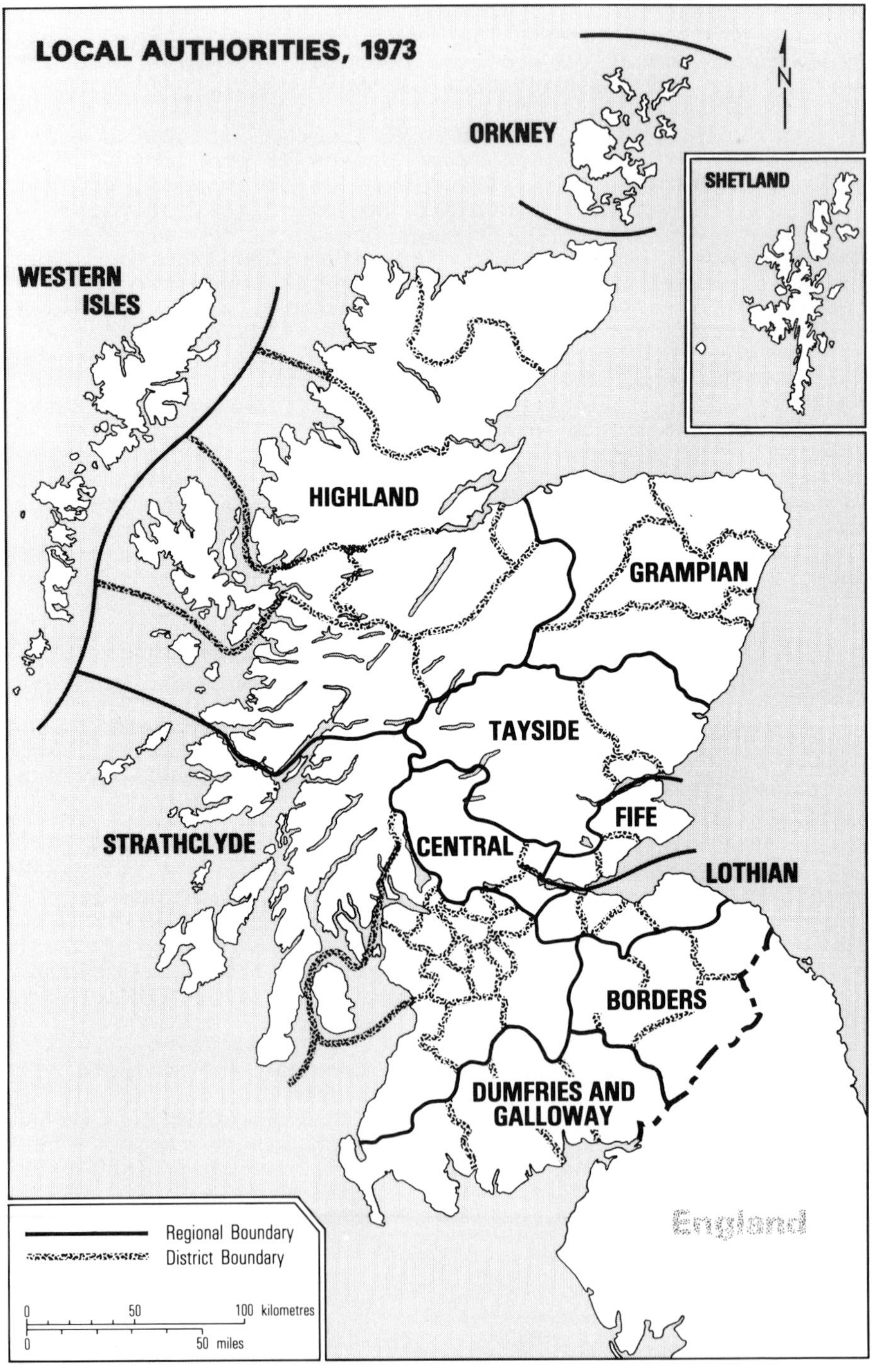
LOCAL AUTHORITIES, 1973
N
ORKNEY
SHETLAND
WESTERN ISLES
HIGHLAND
GRAMPIAN
TAYSIDE
FIFE
CENTRAL
STRATHCLYDE
LOTHIAN
BORDERS
DUMFRIES AND GALLOWAY
England
Regional Boundary
District Boundary
0
50
100 kilometres
0
50 miles

Figure 5.3

was proposed by Dr. J. Dickson Mabon reversing the Government's intention and seeking to continue the system of RPBs over new areas to be designated by the Secretary of State. Several members spoke in support using many of the arguments outlined earlier. In addition, the attention of the House was drawn to the representations of the Clyde and Ayrshire River Purification Boards and the view that a situation where local authority members would have divided loyalties as both pollution prevention authority and sewage treatment authority might not command public confidence (Hansard, First Standing Scottish Committee, 1/5/73, cols. 1546-1548).

Only Mr. Hugh Brown (Glasgow Provan) spoke against the amendment. He said that the pressure to retain the RPBs was coming from vested interests in the boards themselves and a few anglers, and not from electors or ratepayers. He felt the new local authorities would manage the rivers well and reminded his colleagues that industry was also a major polluter. In the middle of the debate, however, the Secretary of State announced that the Government was prepared to bow to the majority feeling of the House and the amendment retaining the River Purification Boards as semi-independent bodies was carried, with only Mr. Brown voting against.

ADMINISTRATIVE ADJUSTMENTS TO THE PATTERN OF RIVER PURIFICATION BOARDS

The campaign was thus successful and the Scottish River Purification Advisory Committee was asked in October, 1973 to consider the future organisation of the boards in the light of local government reform. SRPAC took evidence but there was no unanimity of view. Some wished to see the status quo retained, arguing that the existing boards had established good and close local contacts that were essential in promoting good river conservation. Others favoured a variety of forms of amalgamation so that the larger areas could more easily support the acquisition and staffing of specialist equipment, particularly that associated with marine survey; for it was now clear that new legislation on pollution was imminent. The Control of Pollution Act, 1974, discussed in Chapter Six, was to extend controls over existing discharges to all tidal waters within three miles of the shore.

Local government representatives took the view that the new pattern of boards should match the new pattern of local authorities insofar as this was possible, thus minimising difficulties of representation and finance. They favoured, therefore, seven boards including one board for the Forth Estuary (Lothian, Central and Fife Regions) and a new board for the Highland Region, where the pressures of developments associated with the discovery and development of oil and gas reserves in the North Sea were now felt to justify a full-scale administration for the prevention of water pollution.

SRPAC agreed and recommended the new pattern of boards depicted in Figure 5.4, each board being made up on one-third appointees of the Secretary of State, as before, one-third representatives of the Regional Councils and one-third representatives of the District Councils. As the Regional Councils would be the sewerage and sewage treatment authorities this meant a significant strengthening of the pollution prevention interest; two-thirds of the membership of each board would now have no direct financial interest in the costs of improving sewage treatment works, although of course all would have an interest in the overall level of regional rates which would be required to cover such costs.

THE REGIONAL WATER BOARDS IN THE LIGHT OF A NEW STRUCTURE OF LOCAL GOVERNMENT

Meanwhile, with respect to water supplies, SWAC had been called upon in 1971 to consider the implications of the Wheatley Commission's proposals for the water supply service. The government's White Paper had made it clear that the regional councils were to be formed and the job of the Committee was thus to fit the boundaries of the existing water boards to those of the proposed regional councils (CMD 4583, 1971). Nevertheless, many witnesses took up Wheatley's challenge that the onus of proof lay with those who did not want to see the water service controlled by regional councils and, as a consequence, the debate on the best form of administration was re-opened. Although the positions of most interested parties and their arguments were by now well known, a clear recommendation by this advisory body as to the most appropriate administrative arrangements was believed to be potentially very influential in future policy making.

But it was not to be. The Committee could reach no unanimous decision and the government accepted the view of the minority that 'added areas' could be devised to accommodate the water service within the remit of the regional councils. It accepted the principle of accountability to the electorate and believed that a straightforward transfer of function in this way would create least disturbance to the regional water boards, which were working well (SWAC, 1972).

In contrast, a majority of SWAC now preferred the concept of an *ad hoc* water authority covering the whole of Central Scotland (the area of the Central Scotland Water Development Board), the main benefit of which was seen as the creation of an authority wholly based on the 'source-to-tap' principle of unified control of sources (including the Loch Lomond scheme) and the advantages that would follow for planning future supplies (SWAC, 1972). Major works of multi-regional supply, it was argued, would be much more easily financed under the guidance of one strong authority and the burden would be more evenly spread over standardised charges for such a large area. But this

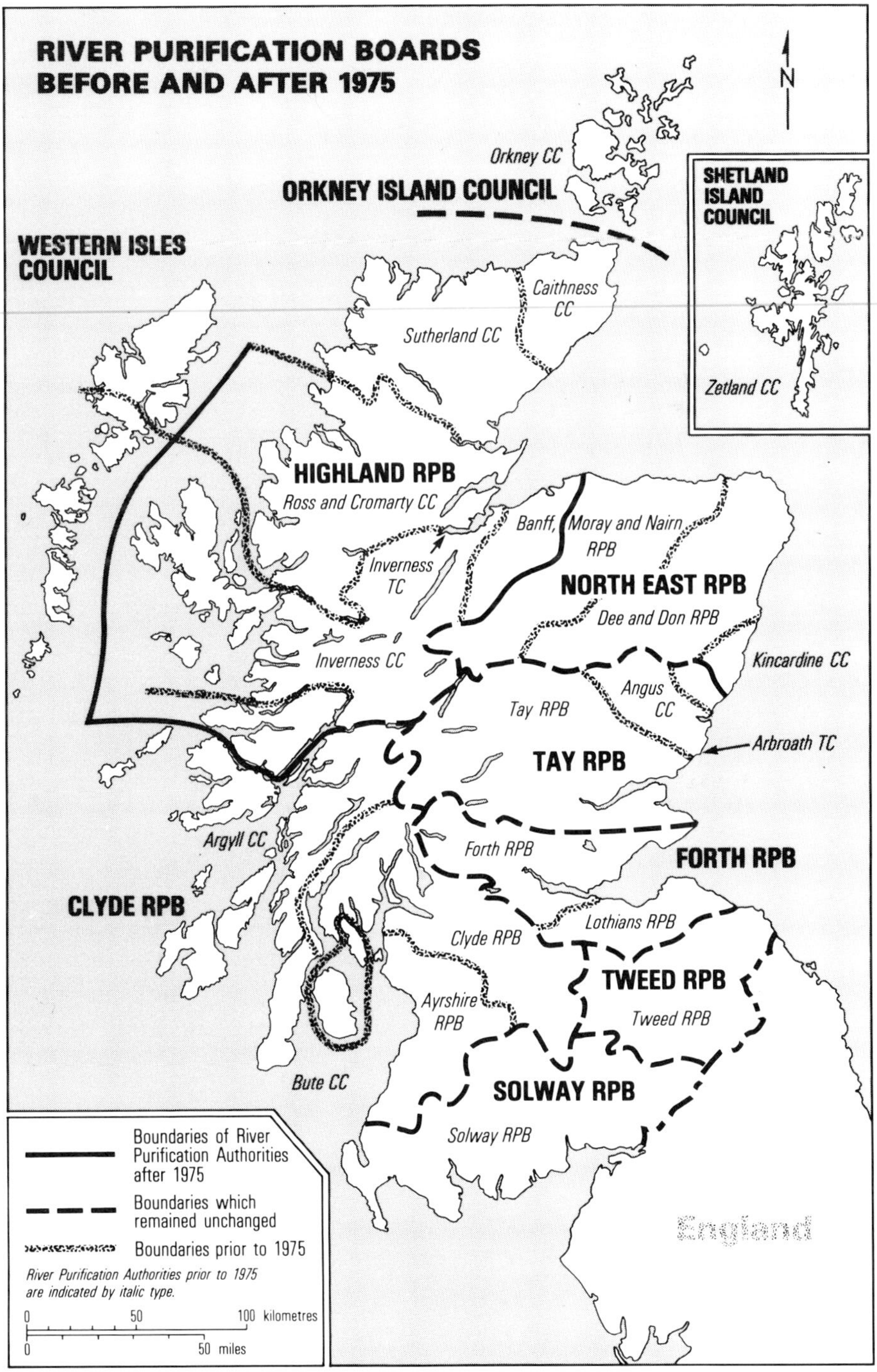
RIVER PURIFICATION BOARDS
BEFORE AND AFTER 1975
N
Orkney CC
ORKNEY ISLAND COUNCIL
SHETLAND
ISLAND
COUNCIL
Zetland CC
WESTERN ISLES
COUNCIL
Caithness
CC
Sutherland CC
HIGHLAND RPB
Ross and Cromarty CC
Banff, Moray and Nairn
RPB
Inverness
TC
NORTH EAST RPB
Dee and Don RPB
Inverness CC
Kincardine CC
Angus
CC
Tay RPB
Arbroath TC
TAY RPB
Argyll CC
Forth RPB
FORTH RPB
CLYDE RPB
Lothians RPB
Clyde RPB
TWEED RPB
Ayrshire
RPB
Tweed RPB
Bute CC
SOLWAY RPB
Solway RPB
England
Boundaries of River Purification Authorities after 1975
Boundaries which remained unchanged
Boundaries prior to 1975
River Purification Authorities prior to 1975 are indicated by italic type.
0
50
100 kilometres
0
50 miles

Figure 5.4

solution would involve indirect representation of 80% of the Scottish population from an area covering about 40% of the country. This, it was argued, was justifiable in this case because water was mainly a technical service, calling for less discussion and decision at the local authority level than most other functions.

Outside the Central Belt the new local authorities would be able to fulfil the role of a water supply agency without any difficulty. The majority expressed their preference for an *ad hoc* authority for Central Scotland on technical grounds, not because they accepted the view that the regional councils would allow the service to suffer in competition for funds with other services. They had not regarded any assumptions about the attitudes of the new regional councils as an important factor in their deliberations.

The argument continued into the House of Commons when the Local Government Bill came to committee with the Government's acceptance of the minority view of SWAC enshrined in the appropriate clause. It was suggested that the Government, having already reversed its policy on the matter of *ad hoc* water authorities with respect to the River Purification Boards, should have no difficulty in doing the same over the matter of Central Scotland's water supply. (Hansard, First Standing Scottish Committee, 8/5/73, Col. 2056-2058).

For the Government, however, Under-Secretary of State Younger replied that the River Purification Boards' functions were regulatory whilst water supply agencies provided a basic service. The existing water boards, he said, spent over £28 million of public money, a substantial sum which should dispose the House towards an administration for water supply that was directly accountable to elected representatives of ratepayers, who provided approximately £16 million, the remainder coming from industrial consumers paying on a volume basis and from specific grants from central government. The Government did not accept the view that any existing *ad hoc* authorities should remain after reorganisation or that any new ones should be created. No mention was made of the same Government's decision to take the entire water industry out of the direct control of locally-elected representatives in England and Wales. (Hansard, First Standing Scottish Committee, 8/5/73, Col. 2068).

Nevertheless, an amendment was moved with a view to giving effect to SWAC's majority view by creating a Central Scotland Water Authority. The argument was conducted in purely Scottish terms, the Government being accused of shutting its eyes to the facts of history. It was argued that very few local authorities had ever been willing to spend the right amount of money on future supplies of water at the right time. There were no votes to be won in spending vast sums of money to ensure that there would be an adequate supply of water in ten or twenty years' time. The water service, if returned to the control of elected councils, would be at the end of the list again and in ten or twenty years' time difficulties would occur.

Under-Secretary of State Younger pointed out that the Government's decision to allow Fife the status of a Regional Council contrary to their earlier intentions and Wheatley's recommendations, had removed one of the most difficult incompatibilities of boundary between the present 'source-to-tap boundaries and those proposed for the new structure of administration. Further, he announced a major new initiative with regard to the planning of future supplies for the whole of Scotland under the auspices of the Scottish Development Department. SDD would shortly publish an important planning document which would include an account of the potential sources available to satisfy an assessment of the likely future needs of different parts of Scotland.

Mr. Younger thought that central government and the new local authorities should jointly plan the future water supply of Scotland because an *ad hoc* authority doing so would be divorced from the strategic plans of the new regions and its decisions would give rise to conflicts, for example, whether priority should be given to the Ayrshire or the Dundee side of Central Scotland. Choices of this kind should not be made by an *ad hoc* organisation solely concerned with technical efficiency and not generally responsible to the electorate for a wider range of strategic considerations and services. Control over water resources and their development should be exercised by the regional council responsible for planning and the strategic growth of the area, or through the Central Scotland Water Development Board which was to be retained in its existing role as a vehicle for consultation and co-operation between regions. In his (Younger's) view, the real choice lay between preserving the existing thirteen Regional Water Boards or making water supplies the direct responsibility of an elected body.

The amendment was defeated by 17 votes to 7. As a consequence of the Local Government (Scotland) Act 1973, Scotland's water supplies were made the responsibility of the nine regional and three island councils, which also undertake strategic planning and provide sewerage, while the prevention of pollution remains the preserve of separate RPBs.

A WIDER ROLE FOR RIVER PURIFICATION BOARDS?

With an eye to developments in England and Wales, where River Authorities had been created in 1963 from the original River Boards, with functions of water conservation as well as quality control, fisheries and drainage management, at least some Scottish River Inspectors were keen to extend their role in water management in the 1960's. The Association of Scottish River Purification Boards and three individual Boards, the Clyde, Tweed and Solway, all made representations to SWAC in 1965, then considering the water supply service of Central Scotland, as discussed in Chapter 3. The Association stressed the role of RPBs in hydrological surveys and in monitoring the quality of water, and that it should be formally recognised by allowing RPB's to be represented on the proposed regional water boards. This view ran counter to SWAC's view that the regional water boards should be composed of representatives of those local authorities which were requisitioned for the board's expenditure and its view that the Secretary of State, through the Scottish Development Department, had overall responsibility for the co-ordination of water management measures around the country. The suggestion was therefore rejected (SWAC, 1966).

The Tweed RPB argued that public water supplies were only one aspect of the development of water resources and that all aspects of river management should be the responsibility of one authority covering entire river basins, although in oral evidence its witnesses agreed that it might be too early to take such a step in Scotland, bearing in mind that 'existing' discharges had yet to come under the control of the consent system. SWAC had not felt qualified to express an opinion on the merits of this proposal, which in effect involved the creation of river authorities similar to those established in England and Wales under the Water Resources Act 1963. It could only comment that the serious competition for limited supplies of water, which was thought to have been a factor in the decision to set up river authorities in England, had no parallel in Scotland where abundant supplies were generally available. The question of creating authorities which would draw together all the interests concerned with water was outside its remit, which primarily embraced the provision of adequate supplies of water to aid other developments, of industry and of housing.

The Clyde RPB pointed out that, in the Clyde Valley, more water for industrial use was drawn from rivers and canals than was supplied by water authorities. The Board thought that RPBs should be consulted about all new abstractions (public and private); unless all abstractions were controlled, the work of local water authorities might well be prejudiced. SWAC responded by pointing out that a proposal to introduce controls over private abstractions of water would change the nature of the reforms it was considering for water supplies from an adjustment of existing arrangements which concerned local authorities only, to a reform which significantly extended the degree of public control over private riparian rights. This might well stir opposition to the extent that the benefits of rationalising the administrative structure of the water supply service might be seriously delayed. As was outlined in Chapter 3, it was envisaged at this time that regional water boards could be fashioned using existing legislation, the 1946 Act, without further reference to Parliament.

A further chance to consider a wider role for the RPBs came with the appointment of the Committee on Scottish Salmon and Trout Fisheries (the Hunter Committee). The Committee's conclusions concerning the administration and management of fisheries were published in 1965 and are considered, with subsequent events in this sphere, in Chapter 6 (CMND 2691, 1965). The Committee had examined the possibility of establishing river authorities in Scotland but did not favour such a change in view of opposition from fishery interests who feared that the establishment of any public agency of control over their private properties would inevitably mean a form of creeping nationalisation. They also thought they would be constantly out-voted on any multi-purpose authority and that other interests would always take priority.

The Hunter Committee firmly recommended that local fishery administration should be separate from other aspects of water management. It believed that the dangers presented to fisheries by pollution, water abstraction and land drainage were less serious in Scotland and thought that it would be a 'long step' from the existing partial coverage of Scotland by District Salmon Fisheries Boards and River Purification Boards to a system of river authorities on the English model. The Committee suggested that the RPB's should use their powers to ensure that all waters which sustained fish life continued to do so, and it expressed the hope that some waters which had ceased to support fish might once again do so.

Nevertheless, the river inspectors' calls for a wider role did not go entirely unheeded, although additional powers came in a very partial manner and through a highly unlikely source of support, the landowners. The relatively new practice of spray irrigation was causing problems in certain agricultural areas on the east coast, particularly in especially dry summers. Some sort of constraint was required on the freedom of landowners to drain local streams for purposes of spray irrigation. The Spray Irrigation (Scotland) Act of 1964 authorised the Secretary of State to make an Order empowering River Purification Boards to license abstractions for this purpose. Such an Order could be made for specific areas at the request of the RPB's which in turn should be acting at the request of the landowners affected. In fact, only one such Order has ever been made, concerning part of East Lothian, although it is said that the existence of formal powers to intervene in the right circumstances currently gives RPBs an authoritative position in the informal resolution of disputes that have arisen from time to time elsewhere.

Abstractions for spray irrigation notwithstanding, the River Inspectors found themselves firmly placed on the margin of Scottish water management in the late 1960s. The reform of local government in general, however, gave them another opportunity to express their views on their future role and indeed on the structure of Scottish water management as a whole.

6

Expanding the range of choice

As the last chapter showed, the path of integration of purposes in water management in Scotland has been towards wider social and economic considerations rather than towards the interrelationships of the hydrological cycle. It follows that any expansion in the range of choice of strategies for the individual functions will be evident within each of the three aspects of water management featured in this book rather than in terms of relationships between functions. In this chapter, therefore, developments in each aspect are discussed in turn.

WATER SUPPLIES

Traditionally, rising demands for water have been met by constructing new facilities. The national survey of resources conducted by the Department of Health for Scotland (DHS) during the 1940s (and considered in Chapter 3) confirmed that there were many potential catchments suitable for impoundment, although their development would best proceed in a regional context rather than through the then-universal local systems. As was shown in that chapter, the regional water boards that existed between 1968 and 1974 were in essence established to develop this potential. Spurred by fears for the future of long-term planning when it became clear that water supply was to return to the full control of local authorities with the reorganisation of local government in Scotland, the Scottish Development Department (SDD) undertook a review of the development of water resources in Scotland, which was published in 1973 and aptly entitled *A Measure of Plenty* (SDD, 1973).

This survey confirmed the predominance of the traditional strategy of building new reservoirs and avoiding, except in a few rural areas, the option of using rivers as sources of supply. In 1971, 91 per cent of public water supplies in Central Scotland (which accounts for 81 per cent of all supplies) came from reservoirs or natural lochs.

In the rest of the country, the proportion was 65 per cent and every regional board obtained over 70 per cent of its supplies from such sources, with the notable exception of North-East Scotland (15 per cent). The average yield of reservoirs was small, reflecting the highly-fragmented administrative structure of the past. Only in North-East Scotland did a significant proportion of supplies (58 per cent) come from river intakes.

The reason for this concentration on impoundment in clear: in the past, sources in the uplands offered the only way of ensuring water of sufficient quality and allowed several major schemes to proceed without the expense of treatment plants. It is significant that much of the water abstracted by the North East Scotland Water Board came from the River Dee which is uniquely protected from pollution because almost its entire upper catchment is either in, or adjacent to, a royal estate, Balmoral. Early schemes of water supply rarely had to go further than twenty miles for a suitable site for a reservoir, but as nearby sources became scarcer, larger authorities went further afield and proposals made by DHS on the basis of the wartime surveys and the creation of regional water boards have facilitated the extension of this trend to other authorities.

The Lothian Region's reservoir in the Megget Valley is typical of this approach. Edinburgh Corporation had gained Parliamentary approval for the development of the adjacent catchments of the Menzion, Fruid and Talla tributaries of the River Tweed as early as 1895 and had been drawing water by aqueduct over the intervening 51 kms (32 miles) from a series of phased developments since 1905. The Megget scheme provides a useful duplication of this trunk main and completes Edinburgh's appropriation of the waters of the upper Tweed, thereby assuring the Lothian Region of supplies well into the next century. Moreover, it is itself capable of several further phases of development. (see Fig. 6.1).

More recently, in unpolluted river basins, largely on the margins of the Highlands, the balance of engineering economics has swung towards schemes of river abstraction as it has elsewhere, especially where a measure of regulation already exists in schemes of hydro-electric development upstream; Fife Regional Council's scheme to tap the resources of the River Earn is one such example.

Such abstractions are, however, confined to rural areas with remote catchments which are unlikely to see polluting developments, and the acquisition by the River Purification Boards of comprehensive powers of control over water quality thus had little significance for water supply. The major development of that time, the Loch Lomond scheme, was in part an extension of the philosophy of tapping large natural reservoirs which had been pioneered in Loch Katrine by the City of Glasgow in 1862. As Chapter 3 showed, what was new about this scheme was not only the scale of regional

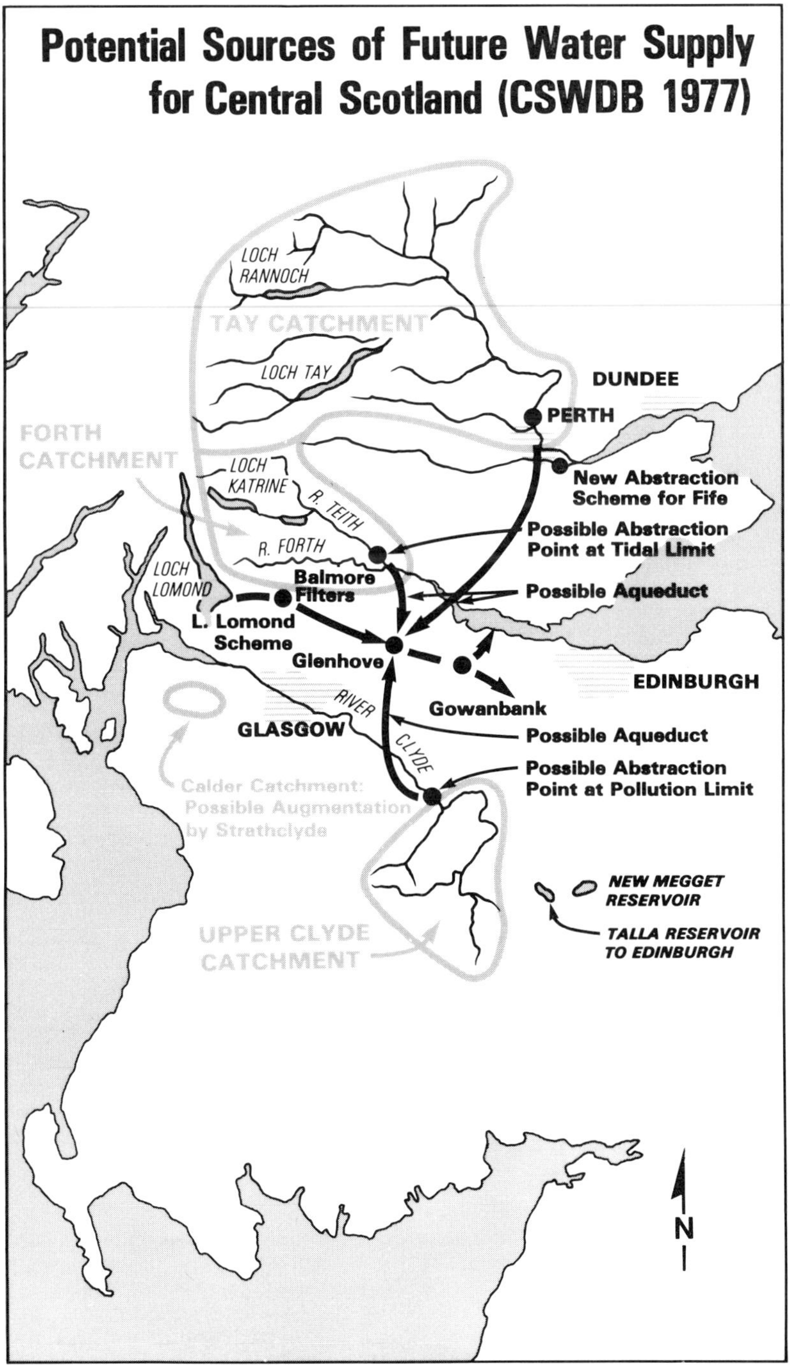
Potential Sources of Future Water Supply for Central Scotland (CSWDB 1977)
LOCH RANNOCH
TAY CATCHMENT
LOCH TAY
DUNDEE
PERTH
FORTH CATCHMENT
LOCH KATRINE
R. TEITH
New Abstraction Scheme for Fife
Possible Abstraction Point at Tidal Limit
R. FORTH
LOCH LOMOND
Balmore Filters
Possible Aqueduct
L. Lomond Scheme
Glenhove
EDINBURGH
Gowanbank
RIVER CLYDE
GLASGOW
Possible Aqueduct
Possible Abstraction Point at Pollution Limit
Calder Catchment: Possible Augmentation by Strathclyde
NEW MEGGET RESERVOIR
TALLA RESERVOIR TO EDINBURGH
UPPER CLYDE CATCHMENT
N

Figure 6.1

co-operation that its financing demanded, but also the extent to which it was implemented ahead of demand, to act as a strategic reserve for major industrial development in Central Scotland and so to serve as a tool of the prevailing strategy for national economic development.

Even though 28 per cent (28 mgd) of the water available through the Loch Lomond scheme remains uncommitted, a second major theme of *A Measure of Plenty* was the identification of its potential successors. The 'Measure of Plenty' of the report's title referred not so much to the existing position as to the range of choice of future options, both locally and of sufficient size to succeed Loch Lomond. The aim of the report was to demonstrate that there was a fairly wide range of practical options, although no attempt was made to evaluate them since SDD regarded this task as the responsibility of the implementing authorities. In its view, the final selection ... "must be left to the water authorities themselves because it is they who are in the best position to make the decision, which will not always rest solely on engineering economics but will need to take account of local amenity considerations including fishing and outdoor recreation interests before the final decision may be reached. In the end it may well prove that because of a failure to resolve a conflict of interest, it will be possible to develop some of the sources to a limited extent for water supply purposes. Indeed, it may not prove feasible to develop some of them at all." (SDD 1973, p. 19)

The report did, however, give some guidance on the strategies of resource development that might usefully be employed for such an inter-regional scheme and on the scale of demand that might be expected to the end of the century and beyond, both in Scotland as a whole and in the regions. Since there were no natural freshwater lochs still to be tapped for Central Scotland's needs, the report supported a strategy of abstraction from rivers, supplemented by regulation of the river flow where necessary. The principal arguments in favour of this strategy included not only such trends as the rising costs of fixed trunk mains (greatly reduced when the rivers themselves were used as aqueducts over substantial distances) but also the greater flexibility of such projects in the face of uncertain levels of demand. The range of potential demand in Central Scotland was thought to be highly uncertain, varying according to different assumptions from an absolute low of an additional 19 mgd in 1991 to a high of 368 mgd in 2031 (SDD 1973, p. 2980, CSWDB 1977, p. 26)

The strategy of river abstraction has two disadvantages firstly, pumping costs are likely to rise over time, thereby increasing the running costs of each scheme; and secondly, there is a greater risk of supplies being contaminated. However, since the River Purification Boards were now in control, this risk need not preclude the adoption of river regulating schemes if the engineering economies prove them to be right. In some catchments further work to regulate the flow would be unnecessary since the construction of hydro-electric schemes had already done so. SDD noted, 'the benefits which could accrue to the water industry by taking advantage of the increased minimum flow so provided', and commented that 'the increasing demands for public water supplies may through time lead to consideration being given to some adjustment of hydro-electric operations to suit the provision of water supplies'.

There were obvious possibilities of joint use for hydro-electric generation and for water supply of reservoirs in the basins of the Tay and the Earn and of Loch Doon (Ayrshire). Conjunctive use of some existing schemes nearest to areas of highest demand seemed a sensible and viable strategy for the future whereby these schemes could be 'overdrawn' well beyond the level of their designed reliable yield in the knowledge that, in a long dry spell, a strategic reserve was on hand from which water supplies could be replenished.

SDD had not entirely dismissed other possibilities, although they were given much less attention. In the report SDD announced its intention of watching closely for any developments in the design of estuarial barrages, although in view of the range of more straightforward alternatives that was available, the application of such a strategy was unlikely. Underground sources, SDD suggested, might also make a useful contribution and it proposed to seek more information on their extent; but it was not expected that they would make a significant contribution to solving problems of future water supply.

In the following year, 1974, the Central Scotland Water Development Board (CSWDB), exercising its responsibility for large, nationally-conceived sources of supply, commissioned an Edinburgh firm of consulting engineers, R. H. Cuthbertson and Partners (hereafter referred to as "Cuthbertsons"), to evaluate the sources of supply that might be considered as major additions once the water from the Loch Lomond scheme had been fully committed. Cuthbertsons showed subsequently that the phasing of river abstraction and regulation schemes could be very much more flexible and hence less demanding financially than traditional schemes of direct supply.

Cuthbertson's report was published in 1977 and is also notable for the extent to which factors other than engineering economics were taken into consideration. The major options, relating to the Clyde, Forth and Tay rivers, were reviewed in terms not only of engineering and financial feasibility but also of their environmental and social impacts. In summary, the consultant's conclusions with respect to each option were as follows:

The Tay Option. Existing water supplies represented an insignificant loss to the catchment and the quality of water was good, the town of Perth having obtained its supply from the Tay for over a hundred years. By the same token, however, the river supported large stocks of salmon and was the subject of intense local interest for both sport angling and commercial netting. Extensive hydro-electric works in the upper catchments since the 1940s had already transformed the flow of the river, the minimum flow of which had increased by 400 mgd. Since the upper target of abstraction for water supply was 200 mgd, it seemed unlikely that the salmon fisheries would be adversely affected (CSWDB, 1977, pp. 65-88).

The Forth Option. The waters of the Forth basin were already extensively used for water supply and over 10 per cent of the mean annual discharge was being transferred across watersheds (mainly to the Clyde Basin and the Glasgow area). Water draining the scenic area known as the Trossachs was of good quality but salmon stocks had been declining for several decades, following growing industrialisation on the shores of the tidal estuary and accompanying deterioration in water quality at tidal phases crucial for the passage of migratory fish. In addition, the maximum requirement for a new scheme of water supply would require some additional storage in the upper basin which might destroy spawning beds; furthermore, the abstraction of clean water before it mixed with the polluted tidal waters might further inhibit seriously the passage of migratory fish (CSWDB, 1977, pp. 95-118).

The Clyde Option. Although migratory fish stocks of the Clyde River Basin had been destroyed a century or more ago by pollution in the upper estuary, the Clyde River Purification Board warned that the abstraction of wholesome water from the unpolluted upper basin would seriously undermine its efforts to restore the waters of the lower basin to a reasonable standard of quality. The Director told the consultants frankly that his board 'would be strongly opposed to any abstraction of this scale, unless more detailed studies subsequently show fears groundless' (CSWDB, 1977, pp. 119-143).

Each of the three options thus presented problems: the Tay option could lead to conflict with the fishery interests, the Forth option involved a similar, but smaller risk to much less valuable fisheries and also had significant implications for the control of pollution, as did the Clyde option. Cuthbertson's desk study was sufficiently detailed in respect of its assessment of engineering and financial feasibility to select and evaluate the options. The results are presented in summary form in Table 6.1.

The high unit cost of the first phase of the Tay scheme arises from the need to construct an aqueduct from Perth to the existing large distribution mains from Loch Lomond. Since the preferred route is largely by tunnel through volcanic rock, it would be economic to construct such a tunnel to match the maximum yield of the scheme rather than merely that of the first phase. The peaking of unit costs of the Forth and Clyde schemes relates in each case to the provision of a storage reservoir for the second phase of development.

Table 6.1 Regulated River Preferred Schemes: comparative annual and unit costs

		Costs			
		Low Target		High Target	
		455 Ml/d (100 mgd)		910 Ml/d (200 mgd)	
Source	Description of Costs	£m	£/Ml	£m	£/Ml
Tay	Annual Cost	13.4		22.1	
	Unit Cost per Ml		81		66
Forth	Annual Cost	11.5		23.4	
	Unit Cost per Ml		71		69
Clyde	Annual Cost	10.8		21.6	
	Unit Cost per Ml		65		65

Source: CSWDB, 1977, p. 186

The table shows that the Clyde scheme is the cheapest for either the low (100 mgd) or high (200 mgd) target of demand. The proximity of the source to centres of population seems to have been the major factor determining the balance of costs.

Cuthbertsons accordingly recommended in January 1977 that the CSWDB should take steps to secure a water order authorising the necessary development in the upper reaches of the Clyde. Since it seemed likely that such supplies would not be required for at least twenty years, they also recommended that an environmental impact assessment should be undertaken.

In November 1977 the CSWDB commissioned environmental assessments for all three preferred sources. Davidson and Robertson, chartered surveyors and land property valuers, were engaged to assess the implications of the three schemes for agriculture and Johnson-Marshall and Associates, planning consultants, were asked to assess their environmental implications, taking account of Davidson and Robertson's conclusions on agriculture.

The latter concluded that, in terms of increasing agricultural impact, the schemes could be ranked as follows: Tay, Forth, Clyde - a finding which accorded with the views of the Department of Agriculture and Fisheries in Scotland which was also consulted in respect of the national agricultural interest (CSWDB, 1979, p. 7). Johnson-Marshall and Associates applied a relatively simple form of environmental impact analysis, involving separate assessments of the impact of each scheme on the visual amenity, the local ecology and the existing land use (including social impacts on local communities.) They also noted the extent to which impacts might be temporary or irreversible and gave some indication of the extent to which impact might be reduced by redesigning the proposed development, although at this stage no account was taken of the costs of so doing. They concluded that the Tay option would be the best choice, largely because the effects of abstraction from such a large river and the works required to convey the abstracted water into the central system are unlikely to produce long-term deleterious effects, whereas, both the Clyde and the Forth, on the other hand, represent major irreversible change to large areas of countryside and to their river systems. Although some benefits, mainly in terms of recreation may accrue to these areas, this does not, in our view, compensate for the disadvantages of these schemes.

With these conclusions in mind, Cuthbertsons re-examine their initial assumptions concerning the demand for a new scheme. By this time, new assessments of likely demand had become available from the new regional councils and revealed that demand for water was likely to grow only outside traditional centres of industrialisation. Strathclyde Regional Council foresaw little growth of demand in the Clyde Basin, the forecast for the whole of the Region for 2011 now being only a further 18 mgd, compared with the estimate of 140 mgd in *A Measure of Plenty*, the figure which had been used in their initial assumptions concerning demand. Meanwhile, developments associated with the discovery and exploitation of oil and gas in the North Sea had indicated that major demands would be more likely to occur in Eastern Scotland; for example, projections provided by Fife Regional Council indicated that it might now require 12 mgd in 2011, compared with the earlier estimate in *A Measure of Plenty* of 4 mgd. It therefore seemed likely that a lower target of 100 mgd would be more appropriate, as would a downward revision of costs for the Tay scheme (since a smaller tunnel would be required). Aggregation of these more thorough and up-to-date projections by the regional councils suggested that Central Scotland as a whole might require an additional 40 mgd by 2011 instead of the 106 mgd that had been forecast previously. Revision of the capital costs to provide this lower maximum yield gave totals for both the Tay and Forth schemes of around £62 million, though the former was nearer the new centres of demand and would therefore have lower running costs. The Clyde option now appeared to be the most expensive, with a capital cost of £72 million.

The Central Scotland Water Development Board is, therefore, expected to choose a Tay Scheme in due course, but economic recession has greatly reduced estimates of rates of growth in the demand for water. As noted earlier, 28 per cent of the full capacity of the Loch Lomond scheme is still uncommitted; but this water appears expensive to the regions because the capital charges allocable to this water have been accumulating since the scheme's inception (a necessary device to reassure the original partners that they would not be forced to pay for somebody else's water). It is therefore understandable that Strathclyde Regional Council should decide to revert to the DHS plan for Renfrewshire and to designate the Calder Valley as its most likely scheme for the future supply for West-Central Scotland rather that to contemplate further involvement with CSWDB schemes. The likelihood of the Tay scheme proceeding in the near future seems remote and it is, in any case, a variant of the Scottish traditional approach, involving a large but remote and sparsely populated catchment that lacks any potential for major industrial development and requires no regulation because it has already been subject to extensive hydro-electric development.

Even though it seems true to say that the range of choice in Scottish water supply has changed little and appears unlikely to do so in the foreseeable future, it would not be correct to conclude that institutional reforms have brought no benefit in the form of a wider range of strategies. The regional water boards, in their seven years of existence, did initiate a pattern of interconnecting mains and procedures for waste detection that has facilitated an increase in the efficiency with which existing resources are deployed. Such measures, the stuff of day-to-day professional water management, have been continued by the regional councils so that, unless some very large new development is proposed, few regions fear difficulties over future supplies. The lead time required for planning and constructing any such development would permit the uncommitted reserves of the Loch Lomond scheme to be deployed either directly or to replace more local sources at present serving other demands.

The great step forward in the management of Scottish water supply over the last thirty years has not therefore been a radical change in tactics but rather the creation of a structure of agencies (the CSWDB and the regions) that permits maximum flexibility where this is required (in Central Scotland). This flexibility has meant that water charges are low, e.g., 10 pence per pound of rateable value (£17 per annum for a small family house) in Glasgow, compared with 22 pence per pound (£40 per annum for a comparable family house) in Leeds (where charges are as much again for sewerage services). As a consequence, and again in contrast to England and Wales, where relatively high water charges are presented independently from local taxes for other services and have provoked a high profile of awareness of the cost of water services, there has been little talk of charging for domestic water by metering. In England and Wales this facility allows the members of a small household, using little water but living in property with a high assessment for water charges, to pay only for what they use although in practice the standing charge for installing and reading the meter has been set at a level that ensures that the average household pays no less in total charge.

WATER QUALITY

At first sight there would seem to be little scope for expanding the range of choice open to River Purification Boards without further legislation. Their duties to consider applications for consent to discharge and to prosecute any who fail to meet the conditions attached to these consents are statutorily defined. Yet the small numbers of prosecutions that have actually occurred, of the order of one or two per annum, and comments published by the Clyde River Purification Board (see below) cast doubt on this simplistic interpretation.

The Clyde River Purification Board has succinctly summarised the situation in its Annual Report, 1981: "During the first decade ... boards were handicapped by insufficient staff, an almost complete lack of chemical or hydrometric data concerning the receiving waters and in particular, inadequate legal powers since the Rivers (Prevention of Pollution) Scotland Act 1951 gave powers to control only new discharges whereas the bulk of pollution problems arose from long standing discharges" (Clyde River Purification Board, Annual Report for 1981, p.5).

By the mid-1960s, however, the Board had built up its inspectorate, laboratory and hydrology staffs and had developed a basic monitoring network of stations measuring river flow and water quality. A very real impetus was then given in the form of the Rivers (Prevention of Pollution) (Scotland) Act 1965. This proved to be a major turning point as it provided the Board with adequate powers to control all discharges to rivers and allowed for the control of the most important tidal waters. Thus, whereas the 1951 Act effectively halted the deterioration of Scotland's rivers, the real improvement of rivers by means of phased, long-term plans became possible only after the 1965 Act.

A major programme of reconstruction followed and the records show that by 1974, in the Clyde catchment alone, some 40 sewerage works had been rebuilt, 17 trade effluent plants built and 14 discharges had ceased by conversion to closed circuit operation. In addition, some 45 major connections had been made to trade effluents (diverting them from the river to sewage treatment works) or to out-dated sewage works. Since 1975 the pace of progress has slowed down as a result of the severe recession but nevertheless a further 12 sewage works and 10 industrial treatment works were completed in the Clyde catchment.

To sum up the situation today, it would not be unreasonable to claim that the present Board and its predecessors have completely reversed the trends of the previous century of neglect and gone almost two-thirds of the way towards a complete restoration of the water quality in the rivers and coastal waters. Much, however, remains to be done and as is repeatedly emphasized in recent reports, 'considerable effort will be required during the present economic climate just to ensure that the present relatively good water quality is never again allowed to deteriorate' (Clyde River Purification Board, Annual Report for 1981, p.5).

This 'effort' will have to continue what has become the standard policy of all boards, that of improvement by persuasion.

The reasons for this universal adoption of informal persuasion are numerous and complex. At first, before the boards acquired full powers under the 1965 Act, they had little choice if progress was to be made: control over new discharges alone could make only a limited impact. The experience of the first few years of trying to make an impression on prevailing levels of pollution without powers over existing discharges seems to have shaped attitudes and styles ever since. A second tendency promoting a cautious approach has been the potential odium of forcing local industries to close prematurely. Some rivers have had to await the demise of their dominating, polluting industry before any radical improvement could take place, as with the rationalisation of the paper industry that once flourished all along the banks of the Rivers Esk and Carron in the Lothian RPB area.

The principal result of such persuasion has been to transfer individual discharges from rivers to the sewerage network and thence to sewage treatment plants. This relatively painless strategy put most of the burden of maintaining river quality onto public funds (the increased quantities of effluent thus arriving at treatment works making their reconstruction necessary if proper standards of effluent were to be maintained); but such a burden cannot be unlimited because of the control over capital expenditure exercised by the Scottish Development Department. In the face of such cash limits, river inspectors have had little choice but to try to exercise influence and to use persuasion. It is true that the RPBs have progressively tightened the screw on polluters since their formation, but there remain considerable differences throughout Scotland in the extent to which the screw needs to be turned further, if at all. The rural boards, such as those for the Tweed, Solway and Tay, appear to have had the least need to apply any sort of pressure to bring about improvements; not only was pollution not a serious problem but there seems to have been a willingness on the part of polluters to co-operate to improve water quality. The policy of the RPBs was thus to persuade those involved (industrialists, farmers and local authorities) by convincing them of the need for, and benefits to be derived from, clean rivers. The nature of the communities in these areas and their size, unity and pre-existing interest in the welfare of 'their' local rivers, encouraged RPBs to adopt a strategy of establishing close personal contacts. The prevailing feeling in RPBs appeared to be that persuasion and communication were vital to success and should take place in an atmosphere of good personal relations.

In the urban-industrial areas, in contrast, matters were not quite so straightforward, particularly in Central Scotland. Most residents do not live near rivers and many settlements suffer severely from the historical legacy of the first industrial revolution. In an industrial community characterized by poor housing, endemic unemployment and a semi-derelict built environment, it is difficult to advocate action to clean up rivers in advance of other remedial measures, especially when restrictions on discharges may endanger, increase the cost of, or otherwise impede the implementation of plans for new housing and new industry. Similar difficulties arise when a mining village is already under the shadow of pit-closure which will bring mass unemployment to the community.

The difference between urban and rural communities over the prevention of pollution may well be that the impetus for improvement in the latter arose internally while in the former it was promoted externally. Members of RPB were faced with a choice. On the one hand, they could assume a passive role, awaiting indications from communities about their perceived priorities, such as the maintenance of employment or the provision of new housing, and weighing these against the need to improve water quality. Often these objectives were in conflict, and where this was so, and especially when funds for amelioration were scarce, development prevailed. It is difficult to justify a claim for an improved environment for fish when that of people remains unacceptable. Alternatively, as in the Lothian and Clyde basins, an RPB could take the view that it was its job to promote river purification and that it should do its utmost to fulfill its statutory duty. An illustration of such a more active stance is provided by the chairman of the Clyde RPB who regretted in 1976 that his board did not become bolder sooner; restraint in threatening legal proceedings, he believed, was all very well, but if the boards were not firm with offenders they would never make any progress. RPBs must be willing to take court action. They must also educate the public and actual and potential polluters to complement their 'bold' attitude. Both the Lothian and Clyde RPBs took a great deal of care over public relations, issuing press statements, giving public lectures and involving schools in a variety of projects.

Despite differences in emphasis between the various boards over how strongly they should pursue long-term goals, the prevailing philisophy in Scotland remains that improvements are most effective if made voluntarily and that, in general, offenders should be prosecuted only as a last resort. All RPBs have been loth to resort to legal action, both because it might prejudice the good relations they were anxious to foster with dischargers and because there is a certain fear of losing face should the case go against them. A further problem is that under Scottish law, proceedings must be promoted through the appropriate procurator fiscal (or public prosecutor), few of whom have had any experience with, or even particular knowledge of, the law relating to river pollution because of the small number of such cases. This unfortunately increases the risk that a case might not be successful and reinforces the inclination of RPBs to avoid court action; in turn, procurators fiscal may be tempted not to proceed with such cases. In 1973 the Crown Office (the central agency overseeing the work of the procurators fiscal) refused to sanction the prosecution of Haddington Town Council which was discharging an effluent of much lower standard than that required by the Lothian RPB but argued that it could not remedy the situation because of controls by central government over capital expenditure. Such a case raises the question whether a local authority which is taking the most effective action possible under the circumstances (including any financial constraints prevailing at the time) can, in fact, be prosecuted (Lothians River Purification Board Annual Report for 1973, p. 25).

Thus the range of choice open to RPBs appears to be very narrow; they must respond to external events and yet serve continually as an institutionalised pressure group in favour of the rivers in their care. The precise way in which this is done is lost in the personal relationships that exist between RPBs, sewerage departments, local industrialists and councillors. Individual personality also appears to be important although it would be invidious to quote specific examples.

Following established practice in England and Wales, SDD has published surveys of the quality of river water at intervals. From the summary tables of these surveys (Table 6.2) and from the annual reports of individual RPBs, it is clear that there has been a steady, if not spectacular improvement.

As might be expected, the surveys reveal a close correlation between the incidence of severe pollution and the traditional areas of concentrated population and industry. Eighty per cent of the Scottish population live in the Central Belt between the Clyde and Forth and Tay estuaries and it is in the Forth, Lothians and Clyde RPB areas that 'substantial water pollution control problems' occur. Elsewhere there are 'effects of considerable local significance' (SDD, 1975, p. 9).

Table 6.2 Chemical classification of Scottish Rivers: poor quality water (B.O.D. greater than 4 mg/ litre) in non-tidal waters surveyed[1]

	1968		1974		1980	
	kms	% of survey	kms	% of survey	kms	% of survey
Highland RPB	not available		0.8	0	11	0.1
North-East RPB[2]	11	1	76	0.9	20	0.2
Tay RPB[3]	8	1	53	0.8	17	0.3
Forth RPB	130	27	227	6.9	236	7.0
Tweed RPB	6	0.2	1	0	1	0
Solway RPB	9	0.2	11	0.4	4	0.1
Clyde RPB[4]	250	22.4	190	2.2	141	1.6

Notes: 1 - total areas (non tidal) surveyed:
1968 5,082 kms
1974 47,314 kms
1980 47,314 kms

2 - not including part of Kincardinshire in 1968

3 - not including County Angus in 1968

4 - not including Argyllshire in 1968

Sources: 1968 SDD 1972, p. 34
1974 and 1980, SDD 1983, Tables 9.1 and 9.2

Although some of the changes between 1968 and later years are illusory, arising from the greater extent, sophistication and professionalism of later surveys, it is clear that the pace of housing and industrial development generally outstripped that of the control of water quality in the late 1960s and early 1970s. Exceptional progress in containing problems of pollution was made in the Clyde area whilst such pressures appear to have been largely absent in the rural, southern border fringe. Since 1974, significant and extensive improvements in water quality have occurred although these have failed to stem an expansion of the length of poor water in the Forth Basin. New pressures also seem to have produced marginal deterioration in the Highland and Solway fringe areas.

This is a good performance, particularly in the Clyde area, despite the difficulty of promoting water quality for largely aesthetic purposes; for river purification lacks the status that water supply enjoys as an economic good and has to rely mainly on angling associations (which display sporadic and varying enthusiasm) as the principal source of external support.

Within the existing legal framework, RPBs appear to have little choice. The Control of Pollution Act 1974, which codified previous legislation, permitted control to be extended to all existing discharges to marine waters (other than those already controlled under specific provisions in the 1965 Act). In addition, RPBs were granted powers to levy charges for the discharge of effluents into rivers and for the costs of works required to lessen the effects of illegal discharges from an identifiable source.

In the former case, this power could be exercised only if the Secretary of State for Scotland agreed to authorise a code of practice by Order, a provision that seems to indicate its status as a reserve power to be applied at some future date (there having been no moves to produce any such code of practice). The Clyde RPB already enjoys powers to undertake restorative works under a local Act passed in 1972 but has not sought to recover costs, even if those responsible could be identified. This experience also tends to support the view that the powers included in the 1974 Act are reserve powers, included to meet the needs of regional water authorities in England and Wales (which are financed independently of local government). In any event, none of the major new powers contained in this Act has yet received the necessary authorisation from ministers. The Act, which dealt with air and noise pollution and the disposal of domestic waste as well as water pollution, was specifically drafted to be brought in section-by-section. Since 1974 the financial and manpower implications of implementing those sections dealing with water quality have discouraged ministers of successive governments from taking action, although the present government has announced its intention of doing so between July 1982 and July 1986.

The Clyde River Purification Board Act of 1972 merits mention in another context. It shows what can be done by an enthusiastic board, and reveals that the pattern of behaviour by RPBs is not as firmly fixed as might at first appear.

The other principal powers conferred by the 1972 Act concern control of sand and gravel workings and the detection and control of the disposal of polluted water underground. Under the general legislation of the 1974 Act, the Secretary of State for Scotland may intervene by Order to declare certain underground waters 'controlled', but only the Clyde RPB has been actively involved in environmental monitoring of this kind, having had to deal with a particularly serious case immediately before it acquired powers under the 1972 Act (Clyde River Purification Board Annual Report for 1972, p. 36). Waste discharged by a distillery into abandoned mine workings eventually broke through to surface drainage, but the distillery refused to grant access for tests to establish whether it was indeed the source of seepage several miles away. This demonstrated the need for powers of inspection and licensing.

The scope for the individualistic approach is also seen in the Clyde RPB's current dispute with the National Coal Board. The accumulation and subsequent discharge into surface drainage of heavily-acidic waters from mines is a problem throughout Central Scotland and has often had damaging consequences for the receiving streams, as with salmon stocks of the River Girvan in Ayrshire. The problem arises after mines have been closed and pumping and treatment are discontinued. Legal history was made at the Appeal Court in Edinburgh on February 1981 when, for the first time, the National Coal Board was found guilty of polluting a watercourse by means of a discharge from an abandoned mine in Ayrshire. The initial prosecution by the procurator fiscal had failed but the Court of Appeal ordered a reconsideration which resulted in conviction, a fine of £250 and, more importantly, a commitment from the National Coal Board to install a treatment plant even though the mine is no longer operational (Clyde River Purification Board Annual Report for 1981, p. 6).

Another source of powers for RPBs has been EEC regulations concerning the quality of seawater used for bathing and the discharge of substances that are dangerous to the marine environment, particularly near shellfish habitats. Again, the Clyde RPB has played a leading role in monitoring the extent to which improvements are required if these Community regulations are to be met (Hammerton, 1978).

In these three instances the approach of the Clyde RPB has been shown to differ from that of other RPBs (though the Forth RPB is also actively involved in marine monitoring with respect to the EEC regulations). These differences in approach are not easy to explain. It is true that the Clyde RPB is the biggest board, monitoring the area that contains over half of the Scottish population, and that the local authorities and Members of Parliament in its area have been predominantly socialist; but it is difficult to avoid the conclusion that personality, particularly of directors and chairmen, has been an important factor.

SEWERAGE AND SEWAGE TREATMENT

The scope for expanding the range of choice has been least in respect of sewerage. Slowly but surely new sewage treatment plants have been built in response to the almost continual demand arising from new developments (at least during the 1950s and 1960s), to pressure from River Purification Boards and to a desire 'to do the right thing' amongst certain individuals in both local and central government. Several regional sewerage schemes, which permitted economies of scale in the disposal of waste whether to the sea following primary treatment only or to rivers and estuaries after full treatment, were built in the Irvine (Ayrshire), Leven and Ore (Fife) and Esk (Midlothian) valleys under the co-ordinating influence of county councils in the 1950s and 1960s and with the benefit of the advice of DHS and the assistance of grants by virtue of the Rural Water and Sewerage Act of 1944. New Towns, particularly Livingston in the Lothians, have also provided the stimulus (and possibility of grant-aid) for large-scale improvements.

With respect to Livingston, a joint committee of West and Midlothian county councils was established in 1962 to examine the implications of the designation of Livingston as a growth area, and (in contrast to the history of policy over water supplies which is described in Chapter Three) agreement on sewerage was reached relatively quickly. Midlothian County Council (which in the 1960s employed the largest number of fully-qualified drainage engineers in Scotland) was to design and supervise the implementation of a suitable plan and the schemes which followed served a much larger area than that of the New Town (Figure 6.2).

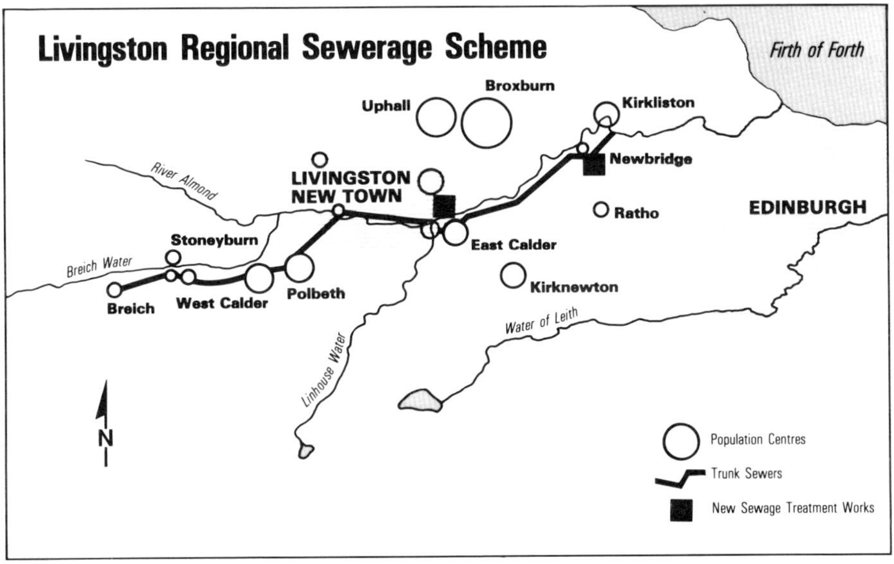

Figure 6.2 Livingston Regional Sewerage Scheme

From its establishment, the New Town Authority wished to make a feature of the River Almond which flowed through the middle of its designated area, and a considerable effort was therefore made to improve the treatment of effluents upstream of the town. A large treatment works now accepts wastes from a wide area by means of an Almond Valley trunk sewer, which accommodates sewage not only from the New Town but also from the industrial villages of West Calder, Addiewell and Stoneyburn, and the communities of the Briech Valley. A second treatment plant nearby at Newbridge serves other industrial centres such as Broxburn, Kirkliston, Newbridge and Ratho to the north of the New Town. These works were all completed by 1969, some time before Livingston's water supply became fully available (Brownlie and Fergusson, 1969).

The fragmented nature of the controlling authorities, which is discussed in Chapters 4 and 5, has confined regional strategies largely to new developments when such an approach is unavoidable and money was available. Coastal works in the appropriate areas followed the extension of the responsibilities of the RPBs to coastal waters and the Cities of Glasgow and Edinburgh both became involved in substantial investments in sewage treatment, in the latter case costing in excess of £11 million (at prices in the early 1970s) for a collector sewer (to replace a series of marine outfalls), primary treatment plant and the dumping of sludge at sea.

The efficiency of sewage works has also been receiving attention. Very frequent operation of the storm overflows installed in systems of sewers or at treatment works reveals that either capacity or design (or both) is inadequate; in short, it is an indicator of the efficiency of the system. Such overflows, which are activated by the overloading of sewers downstream, often result in the excess spilling into surface waters. Attention was drawn to this problem in '*Towards Cleaner Water*'. According to the first edition, 163 unsatisfactory overflows were recorded in 1968, 61 per cent of them on sewerage systems and 39 per cent at treatment plants, indicating undercapacity. By 1975 (the reference date of the second edition), the number of unsatisfactory sewerage overflows had fallen to 87 but the proportion at sewage treatment plants had increased to 45 per cent. A working party was established to examine the problem in detail and this party reported in 1977. The extent of the problem was fully surveyed for the first time and over 1,600 overflows identified. The extent to which each was satisfactory was reported according to the type of controlling authority (that existed in 1974 prior to the reorganization of local government) and no more specific indication was published of the precise location of unsatisfactory overflows. The proportions of overflows reported as either unsatisfactory or unsurveyed, for each category of local authority is shown in Table 6.3.

In all, 1,683 overflows discharged into inland waters and a further 442 into the sea. If it is assumed that attention had not been drawn to the unsatisfactory operation of those recorded as unknown (a status which suggests that day-to-day management is not very active) the problem is most serious in rural areas. These figures reflect an inadequacy of institutional structure for sewerage, with the larger authorities (measured by population and income, i.e., the cities and large burghs) experiencing fewer problems than their weaker counterparts in the rural areas.

Perhaps, significantly, the most recent edition of the rivers pollution survey (*Water Pollution Control in*

Table 6.3 Distribution of unsatisfactory storm overflows by type of authority (1974)

Sewers	% Unsatisfactory	% Unknown
Cities	16	30
Counties	36	25
Large Burghs	26	11
Small Burghs	27	13
Sewage Works		
Cities	0	0
Counties	10	10
Large Burghs	5	11
Small Burghs	16	2

Source: SDD 1977, p. 27

Scotland - Recent Developments, SDD, 1983) discontinues the practice of attempting an assessment of the role of poor sewage treatment in creating unsatisfactory water quality. Perhaps the impact of financial stringency in the current economic climate is reflected in this fact.

Nevertheless, a review of the annual reports of individual boards suggests some considerable progress has been made in the management of sewage treatment since regionalisation of the service in 1975. Generally, closer professional contacts appear to exist between River Purification Boards and regional directorates. Programmes of works appear to have benefited from such co-operation.

Despite the high cost of improving sewerage systems, only a few regional councils have wholeheartedly taken advantage of powers conferred on them by the 1968 Act to charge for the reception of effluents into sewers. Such charges have become a significant source of income, despite the economic recession, to many regional water authorities in England and Wales but are rare in Scotland. Lothian Region took the lead, no doubt spurred by the high cost of recent improvements in sewerage in Edinburgh and other coastal communities, but also reflecting the area's historical distinction in having had more professional staff qualified in sewage treatment than any other local authority, and a history of charging in some parts of the Region before the 1968 Act became effective. The Regional Council was also controlled by an active socialist group for the first two terms of its existence (until 1982).

The scheme of charging adopted by Lothian Region is similar to those used by the regional water authorities in

England and is based on a formula jointly agreed by the (English) National Water Council and the Confederation of British Industry, whereby the charge is composed of four elements, viz., reception costs, the cost of plant and facilities designed on volumetric considerations of biological treatment and of sludge treatment and disposal. Lothian Region had made one change in the practice which is standard in England, in that it has separated capital charges from operating costs with respect to both plant and facilities, thus ensuring that, as a manufacturer's volume of effluent varies with fluctuations in the demand for his products, he nevertheless pays an appropriate charge on the capital equipment that has been reserved for his use at times of full-scale production. It is claimed that this approach is analogous to that whereby a proportion of a manufacturer's electricity bill is attributable to peak demand. A minimum charge for biological treatment both discourages manufacturers from discharging very weak effluents and encourages recycling.

Manufacturers continue to pay the element in general rates that relates to sewerage and sewage treatment to cover the administrative costs and domestic use of the system, and the Regional Council refutes any suggestion that the arrangement involves double charging. Nevertheless, as a gesture of goodwill, those who co-operate with the Council by consistently keeping the quality of their discharge within the consent limits, or by agreeing to discharge at certain times of the day or night only, or by other means, are offered a 20 per cent reduction in the charge. Unlike charges for water, however, these charges are not markedly different from those levied by the regional water authorities in England, for example, 11.59 pence per cubic meter in Lothian (including the reduction for 'good behaviour') compared with 10.33 pence in Yorkshire, 12.43 pence in the Thames Valley and 20.49 in the Northumbrian Region (all charges for 1981-82).

The approach adopted in Lothian Region to the management of trade effluents is very different from that found in most parts of Scotland, where the disposal of sewage and effluents is regarded as a general and public service to be operated on demand and where waste is often accepted without comment to increase the attractiveness of an area for industrial development.

This one exception to the general approach of a constructional response to problems of sewerage and sewage treatment shows that, as was the case with the RPB's in respect of water quality, the range of choice is not as limited as would first appear. There is nothing in the institutional structure that requires a fatalistic approach in the adoption of policies of 'make do and mend', yet these have prevailed except in one authority. Again, satisfactory reasons are difficult to find, but it seems that the roles of political philosophies and of personalities cannot be ignored.

CONCLUSION

Thirty years of water management in Scotland have thus witnessed relatively little change in approach, although the spatial units in terms of which decisions about water management are made have increased significantly in size. This approach differs markedly from that in England and Wales, where the emphasis has been on the whole water cycle, and it can be argued that, by not adopting a holistic approach, Scottish water managers have, in fact, been able to avoid some potential conflict.

The widest range of choice has been considered in respect of water supply, in part because this has been the most important aspect of water management for regional development. Yet even here, the major schemes undertaken during this period have been variants of the traditional approach to securing unpolluted water from the uplands. Other possibilities have been noted, but it is only in respect of prospective developments (and then only if the demand for water increases at the rate it did in the 1960s and early 1970s) that a new approach, of river regulation and abstraction, has been seriously considered on any scale. In large measure, this conservatism of approach reflects the sheer abundance of potential supplies, though the Loch Lomond scheme was new, not only it its scale and finance, but also in the extent to which it anticipated future demand. Expanding the range of choice for those responsible for water quality would require further legislation, though most RPBs have not sought to use all the powers they have and central government has been unwilling to implement legislation already on the statute books because of its financial implications. Instead, RPB's have generally attempted to achieve their objectives by persuasion. In part, this approach reflects the relative unimportance of water pollution, both absolutely in rural areas, and relatively in urban areas (in comparison with other problems facing urban communities), as well as a desire not to imperil new developments that might alleviate these problems. Nevertheless, the example of the Clyde RPB demonstrates that the range of choice has been much wider in practice than the evidence of most RPBs would suggest. The approach to sewerage and sewage treatment has been at an even lower key and has reflected both the low priority attached to this aspect of water management and the availability of finance; even so, the evidence of the Lothian Region indicates that a wider range of choice is possible here also.

7

Retrospect and prospect

The institutional framework for water management in Scotland has undergone some major changes in the past three decades. After almost three-quarters of a century in which the principal laws, policies and administrative structures remained largely unaltered, several important modifications have been introduced. Even so, when measured against the magnitude of changes in other countries in Europe or North America, the pace in Scotland has been relatively modest. The preceding chapters have indicated why this has been the case. As of mid-1982, the scene in Scottish water management seems to be one of consolidation of recent changes, brought about mainly by the reorganization of local government in the mid-1970s, against a background of numerous uncertainties relating to the national economy and government involvement in the water industry.

CONSOLIDATION AND UNCERTAINTY

The Local Government (Scotland) Act of 1973 had two major objectives: to promote more efficient management and to ensure that responsibility for decision-making in a number of specific fields remained in local hands. These objectives were given expression in the establishment of nine regions, each with responsibility for broad structure planning and the provision of certain services, such as housing, education, water supply and sewerage. This meant, notably in the case of water-related services, the transfer of functions from local government agencies or from semi-autonomous bodies, such as the water boards. The years since the 1973 Act have been a period of consolidation and gradual change when the new structures have been made to work. It has also been a period of uncertainty generated by events outside the water industry but with potential implications for it.

The changes in local government enacted in 1973 and implemented in 1975 involved a major upheaval and a new style of local government in Scotland, together with a new planning system. In particular, the attempt has been made to integrate the full range of local government policies through the establishment of policy and resources committees embracing the chairmen of all specialist committees (including those responsible for water supply and sewerage), and through Regional Reports and Structure Plans. As a corollar there has been a considerable strengthening of the central bureaucracy in each region under the control of the chief executive and a relative weakening of specialist departments and their chief officers, including those responsible for water supply and drainage. These developments have affected aspects of water management under the control of the new regional authorities, though the consensus appears to be that the new machinery is working reasonably well.

This major reorganization of local government has also coincided with a period of prolonged and worsening economic depression which would undoubtedly have affected water management even if its organization had remained unchanged. This depression, accompanied by rapid inflation, has placed local government finance under heavy strain as central government has sought to constrain both revenue and capital expenditure by local authorities, the former being subject to indirect influence through the operation of the rate support grant and the latter to direct influence through control of borrowing. The low political priority which water supply and sewerage command has placed these services under severe constraint, though central government has attempted to treat them relatively favourably in respect of capital expenditure. Moreover, there is a degree of momentum about capital spending on projects already under way.

This period of financial stringency has affected not only local government finance but also the economy generally especially in the last three years. Widespread and increasing unemployment, closure of factories and lower rates of economic activity have led to a reduction in the rates of increase in the demand for water, as exemplified by the volume of metered water supplied to non-domestic consumers; factory closures, especially of older plants, have likewise led to some reduction in the volume of pollutants discharged into rivers. Domestic demand has also been affected by the decline in the volume of house building and by the rise in unemployment which have occurred since 1973.

Three other major changes, one actual and two potential, have also had a disturbing effect on government generally and incidentally on water management. The entry of the United Kingdom into the European Economic Community in 1973 has added an additional dimension to government in Scotland through the issue of Community directives, though the implementation of these often depends upon each national government. An increasing number of these directives concerns water management, particularly the control of pollution. More importantly, a considerable degree of uncertainty has been attached to British membership (and hence its consequences) owing to the decision of the Labour

Government, elected in 1974, to undertake a referendum on continued British membership, although, in the event, the proposal to withdraw was conclusively defeated.

A similar degree of uncertainty arose over proposals to devolve powers form the British Parliament in Westminster to a Scottish Assembly. Such an assembly would have had control over natural resources, including water, and of planning, and would have introduced a further level of government. Since one region (Strathclyde) contains two-fifths of the Scottish population, it is unlikely that the regional level of government, introduced by the 1973 Act, would have been retained. Some new structure of water management in Scotland would therefore have been necessary. This issue too was the subject of a referendum, the results from which led the British government to take no further action.

The debate over devolution was, in turn succeeded by a further debate on the relative powers of the regional and district councils, particularly in respect of those functions, such as planning, for which both levels of government have responsibilities, the so-called concurrent functions. The Stodart Committee, which was appointed to advise central government on this issue, was not permitted to consider alternative structures of local government, but had to make recommendations within the framework established in 1973. Although water management was not a concurrent function, it was affected to some extent, particularly by the ability of district councils to commit regional councils to expenditure on water supplies and sewerage for new developments, and the Committee made some minor recommendations affecting water management. The Scottish Association of Directors of Water and Sewerage did, however, use the occasion to voice concern over the disregard by some River Purification Boards of the financial pressures on regional councils and to express in the longer term the preference of most of its members for a national water authority. The Stodart Committee's proposals have largely been embodied in the Local Government and Planning (Scotland) Act 1982.

It is not therefore surprising that the period since 1975 should have been one of slow adjustment in which the new machinery of water management was developed and tested, but in which, against a background of reduced demand and reduced pollution, no major innovation occurred.

CHANGE IN SCOTTISH WATER MANAGEMENT SINCE 1945

Since integrated management has not been a feature of the use of Scottish water, even in respect of the three components on which the analysis has concentrated in this volume - water supply, sewerage and water quality, it is necessary to consider each component separately. Water supply has been the dominant component and has undergone the greatest change. It emerged from the Second World War almost entirely under the control of a large number of local authorities, of very variable size, in which responsibility lay mainly with non-specialist engineers and surveyors. It was already clear that many such authorities were too small and were unlikely to be able to meet the demands of postwar developments and, against the background of the wartime surveys of potential resources (the first such survey ever undertaken in Scotland), attempts were made to encourage the formation of larger boards. Although some joint boards were formed, attempts at voluntary amalgamation were generally unsuccessful, as were attempts to enforce amalgamation by statutory orders. Acceptance by the Scottish Office of the principle of 'source-to-tap' water boards, advocated by both professional and advisory bodies, led to the creation of 13 such water boards outside the structure of local government and in place of more than 200 such local water authorities, under the authority of the Water (Scotland) Act 1967. Although these boards were outside local government, their members comprised elected members of the constituent local authorities and they were financed by precepting these authorities. The creation of these boards was accompanied by the formation of a Central Scotland Water Development Board, covering the territory of seven of them, the purpose of which was to supply water in bulk to two or more of them and to act as an immediately-accessible reserve in times of emergency. The water boards were in turn disbanded in 1975 when water supply became a function of the nine regional authorities formed under the Local Government (Scotland) Act 1973. Water supply then became the responsibility of separate departments in the two most populous regions, but was included with sewerage in joint departments in the remainder. The Central Scotland Water Development Board (CSWDB) was, however, retained with some minor modifications of boundaries, with six regions replacing the boards, though one region, the Borders, subsequently withdrew. The Water (Scotland) Act 1980 was simply an Act to consolidate the various enactments relating to the water supply in Scotland and did not involve any changes.

The control of water quality was also a responsibility of local government in the immediate postwar years, but it had not worked efficiently and was in fact the first aspect of water management to undergo major re-organization. Large areas had been advocated in the 1930s and were endorsed in 1951 by the Broun Lindsey Committee whose recommendations led to the Rivers (Prevention of Pollution) (Scotland) Act. Under this Act nine River Purification Boards were formed between 1954 and 1960, based on river basins and covering most of mainland Scotland except the Highlands. These boards were outside the structure of local government, although two-thirds of their members were elected members of the constituent local authorities and, like the water boards, their finance was obtained by precepting these authorities. Unlike the boards, however, a third of the members of the RPBs were nominated by the Secretary of State for Scotland to represent other interests. The Royal Commission on Local Government in Scotland (the Wheatley Commission) and subsequently the Local Government (Scotland) Bill proposed that, like the water boards, they should be disbanded and their functions transferred to the new regional authorities; but this proposal was resisted and was withdrawn. The number of boards was, however, reduced by amalgamating some of the smaller ones, though a board was also created for the Highlands, to give a total of seven. Although the Forth River Purification Board, which covers the Forth basin, includes most of three regions, the remaining six are broadly coterminous with the regions.

The collection, treatment and/or disposal of sewage have had the simplest history. They were also a responsibility of the local authorities in 1945 and, like water supply, were managed in units of very variable size by non-specialist engineers and surveyors. In contrast, however, no alternative structure was proposed, apart from the suggestion in the 1963 White Paper on Local Government in Scotland that sewerage, with water, should be a responsibility of the top-tier authorities which were proposed. The enactment of the Local Government (Scotland) Act in 1973 thus led in 1975 to the replacement of 234 authorities by departments with responsibility for sewerage in each of the nine regional authorities. This was the last major aspect of Scottish water management to be reorganized and was the principal change affecting water management to be implemented by that Act.

Under the Water (Scotland) Act 1946, the Secretary of State for Scotland was given responsibility for promoting the conservation of water resources in Scotland, the provision of adequate water supplies and the cleanliness of inland and tidal waters, and has exercised these first through the Department of Health for Scotland and then, from 1962, through the Scottish Development Department. Under that Act, a Scottish Water Advisory Committee (SWAC) was appointed to advise the Secretary of State on matters of water supply and a Scottish River Purification Advisory Committee was subsequently formed to provide advice on water quality. Both, however, were abolished in 1982 as part of a general attack on advisory and non-elected bodies.

Other aspects of water management in Scotland are the responsibility of other agencies, though none has comprehensive administrative arrangements akin to those for water supply, sewerage and water quality. The nearest are the District Salmon Fishery Boards, established under the authority of the Salmon Fisheries (Scotland) Act of 1862, which are elected by proprietors of such fisheries, have limited powers to control and improve salmon fishing and cover most of the mainland Scotland outside the industrial areas. The 1976 Salmon and Freshwater Fisheries (Scotland) Act gave powers to the Secretary of State to grant protection in respect of fisheries to whole river systems or substantial parts of them at the instance of angling clubs, though only the Tweed valley has been the subject of such arrangements to date. Other aspects of recreation are covered by the powers given to local authorities under the Countryside (Scotland) Act 1967 to make bye-laws to control the use of rural land and water for recreational purposes. The North of Scotland Hydro Electric Board and the South of Scotland Electricity Board control catchments and lochs used for the generation of hydroelectricity, mainly in the Highlands, and works undertaken in this connection help to regulate the flow of rivers. Land drainage and flood control, which are of minor importance in Scotland, are the responsibility of individual owners of land, though the Secretary of State has powers in respect of large arterial drainage schemes, and regional councils have discretionary powers over flood control on non-agricultural land.

The management of water resources in Scotland is thus complex and has undergone a number of changes in the postwar period. As the subsequent discussion will show, this experience confirms a number of concepts that have been developed elsewhere about the way in which policies and administrative structures are formulated and modified; but the evidence also raises questions about the extent to which local circumstances make generalization difficult.

EXPLAINING THE NATURE AND PACE OF CHANGE

In the main, institutional innovation in Scottish water management since the Second World War has been slow, incremental and often non-sequential, in that steps have been taken in advance of others on which they logically depend, as with the adoption of powers of regulation in advance of the collection of necessary data on pollution. Change has generally affected only one aspect of water management at a time and the three major components of water supply, river purification and sewerage underwent their major postwar modifications at different times (in 1951, 1967 and 1973 respectively). Moreover, it took more than thirty years to give expression to some proposals for the better management of water, even though the problems to which they were response were perceived as urgent at the time. Even when legislation was passed, it was sometimes implemented only slowly (as with the Rivers (Prevention of Pollution) (Scotland) Act in 1951. Changes were generally incremental rather than revolutionary and new policies were largely refinements and improvements of those they replaced. This experience contrasts with that in several other developed countries, such as England and France, where reforms have dealt sequentially with the functions of water management and where subsequent measures have been much bolder and broader in scope, involving comprehensive legislation, integration of functions and a major break with the past.

External Factors

Chapter 1 has shown that there has been a general tendency towards disjointed incrementalism and for change to be effected by both factors external to the water industry and by factors internal to it. Undoubtedly a major external factor in Scottish water management has been the sheer abundance of water in Scotland and the fact that most supplies were obtained cheaply from unpolluted uplands and so required neither pumping nor expensive treatment. Not only was it possible to increase the supply of water by the simple expedient of tapping progressively more distant sources, but there was not much need (at least until the late 1960s) for inter-basin transfers, although Loch Katrine (Forth to Clyde) and Tala reservoir (Tweed to Forth) provided early precedents for such transfers. The abundance and cheapness of water have also helped to keep it low on the political agenda. Shortages of water, although the main cause of complaint by the public, have been only occasional and local. Only when a potential shortage (due to organizational, not natural deficiencies) was seen to threaten regional development in the 1960s did water become a matter of some importance, although a growing concern among both the major political parties over the quality of housing and opportunities for employment has meant that, while water itself was not politically important, major political themes have had increasing significance for water management. The slowness in establishing River Purification Boards in the 1950s was also in part an indication of the low priority given to water quality in a situation where 5 per cent of rivers (by length) were unpolluted and only 8 per cent of public water supplies were obtained from rivers (Scottish Development Department, 1980).

A contributory factor has been the distribution of population in Scotland in relation to its physical geography. Some four-fifths of both population and industrial development are concentrated in the central belt, whereas half the area of the mainland is occupied by the sparsely-populated Highlands. From the viewpoint of water management, Scotland thus falls broadly into three parts: the Highlands, the central belt and the rural south, all with very different balances between demand and supply and very different problems. Once the extension of piped water and sewerage to rural areas and small towns had been achieved in the 1950s, the Highlands and rural south presented few problems for either supply or water quality. The central belt, in contrast, was not only the locus of most of the rising demand for water and sewerage, but also contained the main problem areas and nearly all polluted rivers. Problems of water management and their solutions thus tended to be local and regional rather than national, and this lack of national problems clearly weakened their claims to political attention.

The structure of local government superimposed on that physical base has been another major factor. Indeed, it can be fairly claimed that the reorganization of water management in 1975 was merely an incidental feature of that major change in local government under the 1973 Act. The restructuring of local government in 1929 had tended to set town against country, for although it was widely recognized that many water authorities were too small, the development of better-coordinated systems would generally be to the disadvantage of urban ratepayers whose supplies had been developed earlier at lower cost. This conflict of interest was a major obstacle to the schemes of amalgamation proposed in the 1950s and 1960s and, coming to a head over supplies for Irvine and Livingston New Towns, led directly to the legislation in 1967 to establish water boards. The low priority attaching to water supply and, *a fortiori*, to sewerage also inhibited desirable capital expenditure and the small scale of many authorities made it impossible for them either to employ qualified staff or to find the necessary finance.

The major restructuring of local government in 1975 had its roots in a widespread recognition of the inadequacies of the existing structure generally for a modern Scotland, with particular reference to the needs for regional development and planning. Water was merely one component of the adequate infrastructure that was necessary for such development and was not examined in depth by the Wheatley Commission whose views on the desirable size of units of local government derived from those of planners and others. Water management was seen by the Commission and by the government as a logical function of the upper of the two tiers of local government which it envisaged. That reorganization had to strike a balance between democratic control over policy and expenditure, and efficiency, which often implied large units.

The Commission set its face firmly against both *ad hoc* bodies, which drew their finance by precept from the local authorities, and joint committees. It was therefore logical that it should propose the abolition of both the water boards and the River Purification Boards and that their functions should be given, together with those for sewerage (which had remained within local government throughout), to the new regions. There was, indeed, something schizophrenic about the creation of the water boards while the Wheatley Commission was sitting, since they ran counter to the principle of democratic control; but it was unfortunate from the viewpoint of the boards that they had not been established in sufficient time to influence the debate on their future. In the event it was difficult to argue strongly for their retention, given the preference for democratic control, the alleged interdependence of water and planning, the large size and financial strength of the proposed regions and their broad correspondence with the major hydrological units.

It is true that these points were equally relevant to the River Purification Boards, but these were able to argue that the local authorities were major polluters and could not be trusted to enforce standards; they were also able to muster powerful support and to organize an effective lobby against their proposed abolition. The fears of officials in the water boards that the incorporation of water supply in local government would lead to a weakening of its position (since it would have to compete with all other services for resources in a situation where it carried little political weight) have subsequently been borne out, although it is impossible to know how water boards would have fared, had they survived, in the harsh economic climate since 1975.

The complaint that River Purification Boards have not taken full account of that climate in the demands they have made on local authorities suggests, however, that water supply might have done better financially had it remained with the water boards.

The steadily increasing demand for water, which rose by half between 1950 and 1978, has also been a factor in the greater prominence given to this aspect of water management. This growth was due, not to a larger population (which grew by only 2 per cent) but to rising standards of living and housing and to the redistribution of population. The poor standard of housing in Scotland had long been recognized and its replacement with better-equipped dwellings in major programmes of house building by local and other authorities has been a major reason for the increase in demand. Moreover, that rebuilding often meant a relocation of population in peripheral housing estates and in New and Expanded Towns, and so a need for new supplies and new sewerage. Industrial development also played a part, particularly where industries consuming large quantities of water were involved, as with the petro-chemical plant at Grangemouth. The potential needs of industry were perceived as important, not only as stimuli to the development of new sources of supply, but also because local authorities were loath to place obstacles in the way of new industries or to cause the decline or closure of existing plant by actions that would raise rates.

The demand for water and measures to satisfy it were also affected by the state of the economy. Although demand rose throughout the postwar period, it rose most sharply in the 1960s and early 1970s, which are seen in retrospect as years of relative prosperity when central government was actively promoting economic development in Scotland. The response of the water industry to that demand was in turn influenced by constraints on public expenditure, both in the 1950s and, more especially, in the years since 1975.

A further factor shaping institutional development in the water industry in the postwar period has been the changing role of central government, operating through both primary and secondary (delegated) legislation, through committees of inquiry and advisory bodies, and through the influence of officials on proposals for development. Before the Second World War, the Scottish Office had played only a minor role in water management, being primarily concerned to ensure that minimum standards were maintained, though some developments and proposals did foreshadow postwar developments. In particular, the Department of Health for Scotland had begun to play an active part during the War when it initiated the comprehensive evaluation of Scottish water resources that was to form the basis for most proposals for amalgamation and development in the 1950s and 1960s. The responsibility placed upon the Secretary of State by the 1946 Water (Scotland) Act and the activities of the advisory committees were further indications of a change of emphasis.

It was, however, the increasing political concern with economic development, and hence the need to ensure appropriate infrastructure, that was the major stimulus to involvement by the Scottish Office in Scottish water management, both directly and through the reform of local government. The formation of the Scottish Development Department in 1962 provided a focus for both aspects, since it had responsibility for both water supplies and sewerage on the one hand and for housing, planning and local government on the other. It was the frustration of its attempts to secure more effective management of water supply through voluntary amalgamation and subsequently through orders that led to the passage of the Water (Scotland) Act in 1967. And it was the determination of the Government to secure the reform of local government that led to the rapid implementation of the recommendations of the Wheatley Commission in the Local Government (Scotland) Act 1973 and hence to quite substantial changes in the water industry. On the other hand, the Scottish Office also had to take account of the need to constrain public expenditure, particularly in times of economic difficulty. Two examples of such constraint are provided by the Sewerage (Scotland) Act 1968, which was not fully implemented until 1975, and Part II of the Control of Pollution Act 1974, which still awaits implementation.

The timing of legislation relating to water management has often depended on the vagaries of the Parliamentary timetable, in which Scottish legislation does not command high priority. For example, the enactment of the Rivers (Prevention of Pollution) (Scotland) Act in 1951 was to some extent fortuitous and was due in part to the opportunity for the passage of non-controversial legislation presented by an evenly-balanced Parliament. Its subsequent implementation was slow because there was little enthusiasm for it on the part of the local authorities with which the initiative lay. However, although major changes continue to depend on Parliamentary time, the increasing use of delegated legislation, in which orders are made by civil servants under the authority of Acts of Parliament, has also tended to encourage an incremental approach.

Although water management remains primarily a local responsibility, the Scottish Office has also been able to influence it by providing advice to local authorities and by vetting proposals for grant aid or which involve capital expenditure and hence borrowing, over which central government has control. The scale of government in Scotland has made close and informal contact easy and the fact that most of those with responsibilities for water management at local and national levels are known to each other and can easily discuss matters in person or on the telephone also encourages an incremental approach. The scale of government has also facilitated close relations between departments of central government. The fact that civil servants in all branches of the Scottish Office are servants of the Secretary of State perhaps makes formal links less necessary and facilitates consultation. The dispersal of responsibility for various aspects of water management among a number of government departments and agencies therefore has less serious consequences than in larger countries, the more so as water presents a far less pressing problem in Scotland.

There are several other features of the Scottish scene that help to explain the pace and nature of change in Scottish water management. The status of salmon fisheries as private property, the responsibility of individual owners for land drainage and flood control and the role of procurators fiscal in mounting prosecutions have also contributed to the partial nature of, and cautious approach to, water management in Scotland.

There has been some tendency for public policies and institutions in Scotland to resemble those adopted earlier in England and there is thus an English dimension to Scottish water management, although the differences in both the abundance of water resources and the distribution of population and industry have made this influence much less important than in many other aspects of government. English experience has provided some models for Scottish institutions and some Scottish legislation has followed English precedents. The river boards in England, for example, provided one model for the River Purification Boards, English experience with bye-laws for the control of pollution led to their abandonment in the Rivers (Prevention of Pollution) (Scotland) Act 1965, and that Act was itself based on an English Act of 1961 (the delay in part reflecting the much lesser urgency for action in Scotland). The Wheatley Commission and the Local Government (Scotland) Act of 1973 both had English analogues, though the very different solutions proposed and adopted may reflect the different political composition of local government in the two countries, with conservative local authorities mounting strong and effective opposition to the restructuring of local government proposed in England, whereas the much more radical proposals for Scotland were accepted by central government as a matter for immediate action. In other instances, English precedent has been ignored, most notably in the comprehensive water management outside local government control which was adopted in England in the regional water authorities under the 1973 Water Act.

Internal Features

There have also been several internal features that have influenced the nature and pace of institutional change, though the balance between central and local government discussed earlier may perhaps be included under both heads. Probably the most important factor has been the changing relationships between, and relative influence of, the three groups of actors involved in water management in Scotland, viz., elected members, administrative officials and professional officers.

Water management in Scotland has largely been regarded as a local matter and, apart from the River Purification Boards and, for a short while, the water boards, policy has been determined by elected members of local authorities; even in these *ad hoc* bodies all or most members have been elected representatives from the constituent local authorities and have tended to retain that allegiance. Because water management has not been a politically contentious or important issue and is technically complex, elected members have often played a passive role, accepting the advice of officials on technical matters, though inclined to resist expenditure where their authorities are the paymasters. Since 1975, they have had a wider forum in which they can discuss matters of common interest in the Water and Sewerage Committee of the Convention of Scottish Local Authorities. At the national level, although ministers are responsible for policy, they seem to have played little active part in the development of water management, again primarily because water has not been a contentious issue. Nor has there been much involvement by elected members in the Westminster Parliament when legislation has been considered, although the River Purification Boards had some enthusiastic supporters in both 1951 and 1973.

Professionals, particularly engineers, have played a much more active role than elected members and one that has tended to grow in importance. Before the formation of the water boards and the restructuring of local government, the involvement of professionals varied greatly; only the large authorities could afford specialist staff and much of the day-to-day management was in the hands of surveyors and general engineers, with water engineers brought in as consultants as required. The most explicit example of the role of professionals is, however, provided by the river inspectors, whose professional qualifications were specified by the 1951 Act and who, for the first time, constituted a professional group with a vested interest in improving water quality (although, in the absence of public pressure and political support, they adopted a low-key approach at first and gradually strengthened their professional influence). Water engineers, too, increasingly acquired responsibility for water supply and sewerage/sewage treatment, in terms both of policy and individual proposals, and recruits from local government were also among the growing number of professional staff in the Scottish Office. The heyday of water engineers was undoubtedly the period from 1968 to 1975 when the water boards were responsible for water supply; they could then advise policy makers directly and played a dominant role in the operation of these boards. As was noted earlier, professional engineers in the Scottish Office also played an important, if discrete role, in vetting proposals requiring the approval of central government and acting as a source of advice, particularly before the formation of the water boards allowed the latter to recruit highly-qualified professional staff. They continue to liaise closely with water engineers in the local authorities, though their official role is increasingly concerned with financial rather than technical appraisal. Engineers have thus played the dominant professional role and undoubtedly see themselves as capable of managing most, if not all, aspects of the use of Scottish water.

The contribution of non-technical officials is less easy to identify. It has been most obvious in the River Purification Boards where the clerks have acted as a constraint upon professional officers. Administrators also appear to be dominant as the chief executives in the new regional authorities, where water engineers no longer have direct access to policy makers and are only one among many categories of professionals.

In central government, administrative civil servants have traditionally been regarded as the makers of policy or advisers to policy makers, though professional officers seem to have played a much more important role in respect of water. Administrators have been responsible for both legislation and financial control, although secondary legislation is not the responsibility of professional officers, with administrators concentrating on primary legislation.

There is also little doubt that individuals have played a major role in the evolution of Scottish water management. This might be expected as a corollary of the small scale of government in Scotland and the cozy nature of relations between central and local government; but the form of this inquiry, in which most of the actors are still living and in which discussions have been based on the undertakings of confidentiality, makes it impossible to identify these roles. One example must suffice. There is little doubt that the survival of the River Purification Boards in 1973 was a personal triumph for the then-Inspector of the Clyde River Purification Board, who played a major part in mounting the successful lobby against the proposals to abolish them contained in the Local Government (Scotland) Bill.

Professionals have also been active collectively, both through national professional organizations, such as the British Waterworks Association, the Institution of Water Engineers and Scientists and the Institution of Pollution Control, and through purely Scottish institutions, notably the Scottish Association of Directors of Water and Sewerage Services and its predecessors. These have acted as pressure groups working for larger units of management and greater independence from local government, and the former have provided a forum for those in local and central government. They have pressed for regional water authorities and increasingly for a national water authority, and it seems clear that professional views on the structure of Scottish water management have generally been more radical that what has actually been achieved.

The only other significant pressure groups affecting water management in Scotland have consisted of anglers, particularly proprietors of salmon fisheries. These have provided strong backing for River Purification Boards and contributed to both their creation and their survival. They have, however, resisted the incorporation of fishing into water management generally and, indeed, any general management of fisheries, perhaps for fear of the impact of increasing recreational demand for access to valuable private fishings. By jealously safeguarding salmon fisheries, they have also helped to preserve a wider range of options in unpolluted streams for water supply. Fishery interests have also been included among the members of River Purification Boards nominated by the Secretary of State to represent other interests. They and other independent members have generally been recognized as among the most active and valuable members of these boards (as is the experience in similar situations elsewhere, e.g., in the national park boards and committees in England and Wales).

Other interest groups have been active only occasionally. Somewhat surprisingly, neither amenity or recreational interests (other than angling) have played a major role, perhaps because of the abundance of unpolluted inland water and the easy access to the long coastline. The Royal Commission on Environmental Pollution did, however, contribute to the survival of the River Purification Boards, as did the Confederation of British Industry. In contrast, despite the advocacy of the professional engineering associations, the regional water boards could not be saved from absorption into the new local authorities.

THE ADOPTION OF CONCEPTS

Against this background of the rate and nature of change in Scottish water management, discussion of the five concepts outlined in Chapter 1 will inevitably involve some repetition of points made earlier in different contexts. It is also necessary to repeat the warning in that chapter that these concepts and their developments have not been universally adopted in developed countries; rather, they represent common features of the experience in institutional innovation in water management in the postwar period. Compared with that experience, the degree to which these concepts have been adopted in Scotland has been low, although there is no doubt that major improvements have been made in the management of water resources in Scotland over the past forty years.

Broadening Perspectives

A broadening perspective has undoubtedly been a feature of Scottish water management. Initially, the provision of water and sewerage was advocated solely on grounds of public health and was seen as a purely local responsibility; but increasingly these services have been regarded by central government as part of the essential infrastructure for economic development and for planning. To some degree, this change reflects the increasing range of responsibilities of local government and in part a growing political concern with economic and social inequalities throughout the country; a fear that lack of adequate supplies of water might inhibit economic development underlies many of the changes reviewed in previous chapters. The extension of water supply and sewerage to virtually all rural areas in the 1950s and the policy of regional development are both aspects of this broadening perspective.

A concern with environmental aspects long predates the environmental movement as it is not so much related to conservation of amenity and wildlife as to the prevention of pollution, perhaps because of the highly-localized nature of much water pollution in Scotland. This interest has, however, been reinforced by the growth of that movement and by the entry of the United Kingdom into the European Economic Community, which has issued a number of directives affecting various aspects of water management in the Community. It is perhaps significant that the first environmental impact analysis in water management in Scotland has been undertaken as part of the evaluation of alternative strategies for water supply in Central Scotland in the Cuthbertson Report. Management for the generation of hydroelectricity has also been a feature since 1943, chiefly in the Highlands. With the exception of salmon fisheries (which have been the subject of legislation since the mid-19th century) and, to a lesser extent, other fishing, there has also been little interest in the recreational use of water, though mechanisms exist under the Countryside (Scotland) Act for this to be taken into account. Other aspects, such as drainage and flooding, have received some attention, although they present relatively minor problems.

Integration

There has been very little attempt to integrate functions of water management in Scotland at either local or national level. Even where more than one function occurs within an organization, as water supply and sewerage now do in each region, there seem to be few links between them; indeed, there has been a strong preference among professionals for single-purpose management. Nor has there been any attempt to integrate aspects of water management other than the three with which this book has been primarily concerned; there is little interaction between policies for hydroelectricity generation, angling, other recreation, land drainage and flood control, though they often have implications for each other and for water supply and sewage treatment. The lack of integration is true at both national and local levels; there is no mechanism for bringing all interests together comparable to either the regional water authorities in England or to the National Water Council (though the Government has recently decided to abolish the latter in 1983). Nor is there any attempt to integrate activities, such as data collection and planning, across a range of functions, although the River Purification Boards have gradually acquired prime responsibility for collecting data on the flow and quality of rivers throughout Scotland. There is, however, a number of bodies where informal contact can occur, as in meetings of professional groups and of bodies such as the Scottish Association of Directors of Water and Sewerage Services and the Water and Sewerage Committee of the Convention of Scottish Local Authorities. There is also greater integration between policies for water, particularly those relating to water supply, and policies for economic and social development.

The main reason for this lack of integration is undoubtedly the abundance of unpolluted water in Scotland, so that the demand for water can largely be satisfied at low cost from the uplands and other uses of water can be met without serious conflict of interest. The relative unimportance of flooding and the wealth of opportunity for angling and for other forms of water-based recreation are also factors. The pressure to integrate functions is therefore much less than in countries where polluted rivers are a major source of water and where it is a scarce commodity.

Widening the Range of Choice

There has been very little attempt to widen the range of choice until quite recently, again primarily because of the abundance of unpolluted surface waters. In respect of water supply, the general tendency has been to seek more of the same but further afield, although the formation of the Central Scotland Water Development Board and the Cuthbertson Report are both recognition of the fact that the limit of such an approach has now been reached for Central Scotland. Although the underlying rationale of the Central Scotland Water Development Board was financial, to provide a mechanism for meeting the cost of developing resources of interest to more than one authority, it has certainly widened the range of choice. The options considered in the Cuthbertson Report are concerned with river regulation, an approach that has hitherto been used on only one small river in Scotland (apart from being a by-product of the development of hydroelectricity in the Highlands). Several regional councils, with the encouragement of the Scottish Development Department, are undertaking further exploration of underground sources, particularly in the sandstones that underlie several parts of Scotland. Attention is also directed to the loss of water through leakage, which is estimated to be about a quarter of total consumption (Scottish Development Department, 1980, p. 28).

In the past, sewerage and the discharge of waste water have been considered solely on a local basis, but the reorganization of sewerage services that accompanied the reorganization of local government in 1975 has made such an approach inappropriate. No longer, with the whole of the watercourse lying within the territory of one region, can the sewage from one area be discharged into rivers without regard for the consequences for communities downstream. A growing concern with river pollution has also drawn attention to the need for more adequate sewage treatment and, in respect of coastal waters, for more extended outfalls. The duty placed on local authorities by the 1968 Sewerage (Scotland) Act to accept industrial waste can also be regarded as a widening of the range of choice.

In river purification, the approach adopted has been modified in the light of experience, which has shown that neither absolute standards nor by-laws are appropriate. There has thus been a move towards control through negotiation of individual consents, an approach that has been affected both by the difficulties imposed on prosecution by the Scottish legal system and by an unwillingness to challenge local authorities where inadequate treatment causes pollution or to risk discouraging industrial development by imposing too severe conditions. There has, however, been a move towards more realistic penalties and some recognition on the part of polluters that there may be economic benefits in treating waste.

In fishing, the main objective has been to control poaching of salmon, but there has been a limited attempt to consider the development and management of fishing and to extend the range of concern to trout and other freshwater fish.

Water as an Economic Good

The abundance of water and its cheapness have also acted as disincentives to a more economic approach to water management. Even though a distinction has been made between domestic supplies, which are not metered and are subject to a charge which varies only with the rateable value of the property served, and metered supplies for non-domestic users, which are charged at a flat rate per gallon, there has been a general intention to keep charges low, with water regarded as a public service rather than as a commodity to be sold. Charges for metered supplies broadly reflect the average costs of supplying water and do not discriminate against this class of user. It is, however, difficult to ascertain what the true costs of supplying water are and the high initial costs of installing meters in domestic premises has been categorized as a nonsense in a country where water is so abundant. A further problem is the lack of control over the abstraction of water (except that used for spray irrigation).

The costs of sewerage are also a charge on the rates and the low political significance attached to sewerage provides a similar incentive to keep charges as low as possible, a recognition of the adage that there are no votes in sewage. The 1968 Sewerage (Scotland) Act empowered local authorities to make special charges for trade effluent discharged into public sewers, but few have yet elected to do so. River Purification Boards have no powers to charge directly in respect of discharges to rivers, though they will acquire such a power when Part II of the Control of

Pollution Act 1974 is implemented. The boards have preferred to work through consents and have regarded legal action as a last resort (in which the final decision to prosecute does not rest with them). In any case, until recently the level of fines for such breaches of consents has been very low.

Both water boards (when they existed) and River Purification Boards have been resented by local authorities because of their powers to meet their expenditure by levies on the local authorities which the latter must meet, though the boards point out that these charges represent only a very small part of the total expenditure by the local authorities. The latter allege that the boards have not taken sufficient account of the adverse economic climate in which the authorities are operating in the financial demands they make on the authorities. It should, however, be noted that only a small part of the charges for both water and sewerage bear directly on rate payers, since some four-fifths of all expenditure by local authorities are met by central government through the rate support grant. The same is true of the small part of water charges that are levied on both domestic and other consumers as the public water rate.

There is increasing recognition that the costs of providing additional water will be higher in real terms than in the past. Replacing 'ageing assets', i.e., water mains and sewers installed in the 19th century, will also be a burden that users will have to bear (though there is some belief that the scale of the problem has been exaggerated, possibly by those who have a vested interest in replacement). There is however, little indication that charges for water services might be used as a tool of management to affect the demand for such services or to discriminate between different classes of users.

Public Participation

There has also been little tendency for greater public participation in decisions about water management in Scotland. In large measure, this has been due to the fact that most aspects of water use in Scotland have been public services under political control. Sewerage and sewage treatment have always been the responsibility of local authorities and water supply has likewise been for most of the period since the end of the 19th century (apart from the seven years when the water boards were responsible). Under such circumstances, policy has been determined by elected members who regard themselves as representing the interests of their electorates. In that sense, public participation is unnecessary.

It is true that there has been an increasing commitment to public participation in planning, itself a local authority responsibility, and there is provision for public enquiries in respect of proposals for major developments of water services, though few have in fact been held. Water services are considered in the local authorities' structure plans which are themselves the subject of public participation; but the level of public interest in such broad-ranging documents has generally been low and the contribution of water services small, reflecting their low political priority. No proposals for supply reservoirs in Scotland have attracted great public interest and opposition on amenity grounds, as has happened in national parks in England and Wales, although some hydro-electric developments in the Highlands have done so, as has a proposed pump storage scheme near Loch Lomond. It should, of course, be noted that, under the Hydro-Electric Development (Scotland) Act of 1943, the North of Scotland Hydro-Electric Board established advisory committees on amenity and fisheries, though the former has recently been abolished. The Secretary of State also nominates one third of the members of River Purification Boards to represent wider interests and there is some indication that these boards are sensitive to the needs for good public relations and are keen to seek public support for their actions.

Nevertheless, public participation in water management has not, to any large extent, been an issue in Scotland, mainly because of the abundance of water, its low political standing and the fact that it is wholly or largely a public service under the control of elected members. That control is local and that beautiful scenery is not a scarce resource in Scotland may also help to explain the lack of pressure for public participation, since fears of loss of local control and loss of amenity have been major stimuli to demands for public participation elsewhere.

RELATED DEVELOPMENTS

Four developments related to the adoption of these concepts were also identified in Chapter 1, viz., increasing government involvement, river basins as management units, a broadening range of specialists and increasing sophistication of planning, and these have played an important part in water management in Scotland, though the emphasis has varied.

Increasing Government Involvement

Despite its low political priority and its status, for the most part, as a public service, water has nevertheless become a topic of increasing importance in Scotland. At a local level, local government has become increasingly powerful and water services have become a responsibility of the top-tier authorities. These are sufficiently large and wealthy to have separate departments responsible for such services, committees of elected members to determine policies for them and highly-qualified professional staff, a far cry from the situation that existed in most authorities in the 1930s. Water supply and sewerage are recognized as part of the essential infrastructure for new residential and industrial developments and play a part in structure plans. Since

1967, local authorities have also had powers to facilitate the greater recreational use of water, although this remains a minor issue.

At a national level, the Scottish Office has also come to play a more important role, though it is possible to detect a lessening of detailed control with the creation of more powerful regional authorities in 1975 and the tendency to abolish advisory bodies since 1979. Before the Second World War, central government was primarily concerned to ensure minimum standards in water supply and sewerage in the interests of public health and in the construction of reservoirs in the interests of public safety, although there were already indications of a concern with the role of water as part of the necessary infrastructure of industrial development. Water was regarded as a local responsibility, though there was recognition in the 1930s of the need for a better local structure for both water supply and water quality. The wartime surveys of water resources undertaken by the Department of Health for Scotland provided a basis for advice from central government and the 1946 Water (Scotland) Act placed a statutory duty on the Secretary of State to promote the conservation of water resources and the provision of adequate water supplies throughout Scotland. It also authorized the appointment of the Scottish Water Advisory Committee, which was to play an important part. The Rivers (Prevention of Pollution) (Scotland) Act of 1951 similarly led to the appointment of the Scottish River Purification Advisory Committee. The Hydro-Electric Development (Scotland) Act of 1943 had earlier established the North of Scotland Hydro-Electric Board and given it powers to make use of water for the generation of electricity.

The extent of national involvement has fluctuated with the political philosophy of the government in power, its perception of problems and the state of the economy, as well as with the strength of local government, but there has undoubtedly been a wider involvement by government, both at an advisory and consultative level and in terms of a widening range of legislation. Before the restructuring of local government in 1975 (or, in respect of water supply, from the formation of the water boards in 1967), professional staff in central government had come to play an increasingly important role in providing advice to local authorities and in helping to shape Scottish water management through such advice and through technical appraisal of proposals requiring grant-aid or capital expenditure. A more positive approach towards water management (especially of water supply) is evident from the late 1950s, and particularly from the formation of the Scottish Development Department in 1962 and the development of regional economic policy in the 1960s. Not only was there recognition of the place of water in urban and industrial development, but fear of the possible consequences for such development or failure to secure the rationalization of water supply through voluntary amalgamation also led to the creation of the water boards.

The Secretary of State's general responsibility for Scottish water resulted in legislation in the 1950s and 1960s relating to both water supply and water quality, and a number of other measures extended public involvement in other aspects of water management, though largely by passive or delegated powers to local authorities, e.g. land drainage (1958), flood prevention (1961), spray irrigation (1964), recreation (1967) and angling (1976). Directives of the European Economic Community have also placed increasing obligations on central government, particularly in respect of water pollution. The obligation imposed by the 1946 Water (Scotland) Act on the Secretary of State to collect and publish statistics and information does not, however, seem to have received much public expression until the appearance of appraisals of both potential water supply and water quality in respectively *A Measure of Plenty* (Scottish Development Department, 1971) and *Towards Cleaner Water* (Scottish Development Department, 1975). There is also a noticeable gap in the absence of any body, comparable to the (now to be abolished) National Water Council in England, to take an overview of water management in Scotland, and responsibilities remain dispersed among a number of departments and agencies of central government.

Since the reform of local government, central government has tended to play a less active role in view of the expertise now available in the regional authorities and the adoption of a policy of giving local authorities greater freedom to determine spending within approved levels of total expenditure (though there remains tight control over this level and over borrowing for capital expenditure).

River Basins as Management Units

There has been a general trend towards the use of larger units in Scottish water management, though only in part do these correspond with river basins. That tendency began first and is best developed in the control of river pollution with the transfer of responsibility from numerous local authorities to the nine River Purification Boards created in mainland Scotland (except the Highlands) under the authority of the Rivers (Prevention of Pollution) (Scotland) Act 1951. Yet even these areas did not exactly correspond with river basins, nor, given the hydrography of Scotland and the aim of broadly similar areas, could they do so; the areas of the Solway Board, for example, comprises the basins of several small, parallel rivers that flow into the Solway. For extraneous reasons (to avoid the need for regions to be represented on more than one board), the nine boards were reduced to six following the reorganization of local government, though a seventh was added to cover the Highlands. This change made for a poorer fit in three (Clyde, North East and Tay) and a better fit in one (Forth). Only two of these seven, Forth and Tweed, correspond with major river basins, the remainder embracing several basins (with the Clyde and Tay wholly in one board area). Units of water supply likewise experienced a marked increase in size, following the passage of the Water (Scotland) Act 1967, which created thirteen water boards in place of over two hundred water authorities. These water boards, too, had boundaries that were primarily hydrological and showed a general correspondence with river basins; but they were soon replaced by departments of the nine regional councils created under the Local Government (Scotland) Act 1973. Although the territories of these councils were chosed on grounds of general administrative suitability, there was a broad, if fortuitous, correspondence between them and the major hydrological units. Adjustment was further eased by the device of 'added areas', whereby two areas (in Strathclyde and Tayside), which made better sense hydrologically in other regions, were administered for purposes of water supply by those regions. Had the government accepted the recommendations of the Wheatley Commission and not bowed to political pressure, there would, in fact, have been six regions and hence somewhat greater discrepancies. It is also ironic to note that the discordance between River Purification Boards and the regions with which they deal is rather greater than that between regions and hydrological units for water supply.

Trends in respect of sewerage are very similar, except that there was no intermediate stage of large units. There was, in fact, virtually no change until 1975, when the 234 existing authorities were replaced by nine regional departments. As with water supply, there is broad correspondence with the major hydrological units, though the discordance of boundaries is less significant and it has not been necessary to use the device of added areas.

In respect of water supply, professionals have favoured the creation of a national body, and the Central Scotland Water Development Board established under the 1967 Act to provide bulk supplies for two or more of the seven boards that comprise it and the only board of its kind in the United Kingdom, might be regarded as a step in that direction. Since the reform of local government, the CSWDB has comprised the territory of five of the regions. It should also be noted that, for operational purposes, both water supply and sewerage are managed by operational divisions which maintain many affiliations with the smaller units the regions had replaced.

Thus has been a general (if sometimes fortuitous) adoption of river basins, as the bases for units of management for individual water purposes in Scotland. In no sense, however, has the river basin become the basis for integrated management as it has elsewhere.

Broadening the Range of Specialists

Water management in Scotland has, in terms of professional advice, within the public sector and from consultants, been dominated by the engineering profession, though there has been a progressive improvement in the level of expertise available generally. Only the larger local authorities could afford professional water engineers in the 1950s and 1960s, but the creation of the regional water boards provided an opportunity for water engineers to play the dominant role in water supply everywhere. There was similarly little opportunity for professional management of sewerage and sewage treatment until the restructuring of local government created large departments in the new regions; here too, water engineers have played the dominant role. In central government, the Chief Engineer in the Scottish Development Department is the principal professional officer with responsibility for advice on water supply and sewerage, and senior engineers in that department have often had experience in local government.

The main exception to the dominance of the engineering profession has been in river purification, where the task formerly undertaken on a part-time basis by sanitary engineers became the responsibility of chemists and biologists appointed as river inspectors under the 1951 Rivers (Prevention of Pollution) (Scotland) Act. The range of expertise on their staffs has become much wider than that in the local authority departments responsible for water supply and sewerage, and now includes hydrologists. There is, however, no official with responsibility for water quality at a national level corresponding to the Chief Engineer, although the Department of Agriculture and Fisheries for Scotland, which is responsible for fishery policy in Scotland, has a staff of biologists, including a Chief Fisheries Officer. Of course, engineers in the Scottish Development Department have access to professional advice from chemists with responsibilities for pollution generally, to biologists, economists and other professional staff, and more recently, hydrogeologists. Similarly, the fact that water supply and sewerage are a function of local government gives water engineers access to a range of skills in local government, such as chemists, planners and statisticians, though none in an expert in the water industry. In any case, such advice has to be sought and the responsibility for advice and decisions on water policy rests primarily with engineers.

Increasing Sophistication of Planning

It is not easy from the available evidence to be dogmatic about the extent to which planning in water management has become increasingly sophisticated, although the general conclusion is that it has. To a large degree, it is a necessary corollary of the increasing professionalism of Scottish water management and the progressive enlargement of the units of management discussed earlier. At a national level, the wartime surveys and *A Measure of Plenty* provided a perspective for planning water management and have affected the choices of water authorities. Similarly, successive editions of *Towards Cleaner Water* provide a wider perspective on water quality. The location of water supply and sewerage within the new local authorities has also provided access to information necessary for the making of projections and to staff responsible for both physical and policy planning who have themselves adopted increasingly sophisticated methods in formulating their plans. The same is true at a national level. The consultants who have made plans for water authorities in the past have also adopted more sophisticated methods, as the Cuthbertson Report, *Water for Central Scotland,* commissioned by the Central Scotland Water Development Board, shows.

Nonetheless, as the preceding section has demonstrated, desirable skills (biological, chemical, economic, mathematical, statistical) are often lacking in the organization in which plans are made and, despite the improvement in the availability of data, there are notable gaps, in particular on the volume of abstractions and on the true costs of water. A national plan for water (as opposed to national appraisals of potential supplies and water quality) is also lacking.

FUTURE PROSPECTS

As the introduction to this chapter has shown, prediction of future demands and of the future shape of water management in Scotland is particularly difficult because of uncertainties about the context in which the water industry will operate. Demand depends in part upon the size of the population and the state of the economy, and population projections have been revised downwards since *A Measure of Plenty,* and prospects of economic recovery remain gloomy. In any case, following the completion of a number of major projects, there seems little likelihood that further major developments in water supply will be necessary during the remainder of this century; indeed, the fear has been expressed that a future generation of water engineers seeking to gain experience of major constructions will have to go elsewhere. In any case, there remains the uncommitted water from Loch Lomond, although some financial solution will have to be found to make this attractive to water authorities. Uncertainty also exists about the possible financial implications of replacing ageing assets and the extent to which future needs can be met economically by reducing leakage. Some better understanding of the reasons for the high per capita consumption of water is also necessary, and both levels of consumption and future supplies could be affected by the application of economic principles to water management.

Uncertainty also remains over the wider context of water management. The European Economic Community has become increasingly involved in policies that affect water management, especially relating to the control of pollution, and these have implications for both manufacturing industry and the local authorities, and the Government, while wishing to safeguard existing improvements in water quality, is anxious to avoid additional expenditure. An additional complication is that the Labour Party has adopted withdrawal from the Community as an object of policy which it will attempt to implement if elected to office. It is also committed to devolution and the establishment of a Scottish Assembly, as are three of the other parties, a development with considerable implications for the future of Scottish water management, since it might lead to the establishment of a Scottish National Water Authority, a development that by many professional water managers in Scotland would favour on technical grounds.

Water engineers and other in Scottish local authorities have also expressed some reservations about the continued existence of river purification outside local government, and a decision to incorporate that function within the local authorities could be seen as making for greater integration of water management within a single administrative framework. Even so, many aspects of water management would remain the responsibility of other bodies, the most obvious being fisheries. Opposition from proprietors of salmon fisheries has frustrated proposals for more integrated management of freshwater fisheries in Scotland, and it is significant that the Hunter Committee, at the instigation of the River Purification Boards, investigated (but rejected) a proposal that the latter should be given responsibility for fisheries, a proposal that reflected the interdependence of water quality and fisheries and the availability of biologists within the boards. Larger units of management and greater integration of fisheries would receive professional support, and some kind of national water authority (or alternatively, larger *ad hoc* authorities) would seem a logical consequence of any decision to establish a Scottish Assembly.

CONCLUSION

There have been significant changes in water management in Scotland over the past forty years, which have seen the emergence of a more rational and professionally-managed industry. The two major components, water supply and sewerage, remain (as they began) the responsibility of local government and it can be argued that the most significant developments are largely a by-product of the reform of local government in which considerations of water management have played only a minor part. That claim would be stronger if the River Purification Boards had not succeeded in overturning the proposed inclusion of river purification in the new regions. Many minor aspects of water management (judged by their importance in a Scottish context) remain outside local government, whether in private hands or the responsibility of other agencies, and no comprehensive perspective on, or forum for, water exists at either national or local level. There also remains a number of gaps, notably the lack of information about, and control of, abstraction.

The sequence of changes which have been examined in this book shows some similarity to events in other developed countries; but the concepts which have shaped developments elsewhere have generally played a much smaller part in Scottish water management. The main reason is the sheer abundance of water, from which flow the high level of consumption, the tradition of cheap water and its low priority in political decisions. It is true that Scottish geography, Scottish law and Scottish local government have all affected the way in which Scottish water management has evolved, and that economics will play a larger part in the future. But a country in which the total quantity of water consumed is perhaps only a hundredth of that potentially available must be grateful that, 'nature has provided for the water needs of Scotland on a lavish scale'.

Perhaps Scots should also be grateful that the scale of government in Scotland permits a closeness of relationships, a cosiness, both within central government and between centre and region, that makes it relatively easy to resolve difficulties. In that sense, despite this increase in the size of the unit, it can be said, following Schumacher, that 'small is beautiful'.

Bibliography

Abercrombie, Sir Patrick and Matthew, R.H. (1948). *The Clyde Valley Regional Plan.* London: HMSO.

Adams, I.H. (1978). *The making of urban Scotland.* London: Croom Helm.

Baker, W.A.R. (1956). The Caithness Regional Water Supply Scheme. *Proceedings of the Institution of Civil Engineers,* **5**, 366-383.

Beveridge, Sir William, (1942). *Social Insurance and allied services.* CMD 6404. London: HMSO.

Bollens, J.C. and Schmandt, H.J. (1965). *Metropolis: its people, politics and economic life.* London: Harper and Row.

Bower, B.T., Barré, R., Kuhner, J. and Russell, C.S. (1981). *Incentives in water quality management: France and the Ruhr area.* Washington, DC: Resources for the Future, Inc.

Braybrooke, D. and Lindblom, C.E. (1970). *A strategy of decision.* New York; The Free Press.

Bruce, J.P. (1976). The National Flood Damage Reduction Program. *Canadian Water Resources Journal,* **1**(1), 5-14.

Burns, J.O.O. (1952). Recent developments in rural water supplies and sewerage in Scotland. *Civil Engineering and Public Works Review,* **47**, 574-576.

Central Advisory Water Committee (1960). *Final Report of the Trade Effluents Sub-Committee.* Edinburgh: HMSO.

Central Scotland Water Development Board (1977). *Water for Central Scotland.* Edinburgh: HMSO.

Central Scotland Water Development Board (1979). *The way ahead.* Edinburgh: HMSO.

Clyde River Purification Board. *Annual Reports.* Glasgow.

Commissioner for Special Areas (Scotland) (1939). *Report 1938-9.* CMD 5905. London: HMSO.

Cormie, W.H. (1970). The Loch Lomond Water Scheme. *Journal of the Institution of Water Engineers*, **24**, 291-295.

Craine, L.E. (1969). *Water Management Innovations in England*. Baltimore, MD: John Hopkins Press.

Davis, R.K. (1971). *The Range of Choice in Water Management: A Study of Dissolved Oxygen in the Potomac Estuary*. Baltimore, MD: Johns Hopkins Press.

Day, J.C. (1974). Benefit-cost analysis and multiple purpose reservoirs: re-assessment of the conservation authorities Deer Creek Project, Ontario. In: *Priorities in Water Management*, ed. F.M. Leversedge, Victoria, BC: University of Victoria, Department of Geography, Western Geographical Series, **8**, 23-36.

Department of Health for Scotland (1931). *First Report of the Scottish Advisory Committee on Rivers Pollution Prevention*. London: HMSO.

Department of Health for Scotland (1934a). *Annual Report*. CMD 4837. London: HMSO.

Department of Health for Scotland, Committee on Scottish Health Services (1934b). *Interim Report: Water Supplies*. London: HMSO.

Department of Health for Scotland (1935). *Annual Report*. CMD 5123. London: HMSO.

Department of Health for Scotland, Committee on Scottish Health Services (1936a). *Report*. CMD 5204. London: HMSO.

Department of Health for Scotland, Scottish Advisory Committee on River Pollution Prevention (1936b). *Sixth Report*. London: HMSO.

Department of Health for Scotland (with Ministry of Health and Ministry of Agriculture and Fisheries) (1944). *A National Water Policy*. CMD 6515. London: HMSO.

Department of Health for Scotland, Committee on Water Rating in Scotland (1946). *Report on Water Rates and Charges*. CMD 6765. London: HMSO.

Department of Health for Scotland (1948). *Annual Report*. CMD 7659. London: HMSO.

Department of Health for Scotland (1950a). *Annual Report*. London: HMSO.

Department of Health for Scotland, Scottish Water Advisory Committee, River Pollution Sub-Committee (1950b). *Report on Prevention of Pollution of Rivers and Other Waters*. London; HMSO.

Department of Health for Scotland, Drainage of Trade Premises Committee (1954). *Report on Drainage of Trade Premises*. CMD 9117. London: HMSO.

Department of Health for Scotland (1957). *Annual Report*. CMND 385. London: HMSO.

Department of Health for Scotland (1961). *Annual Report.* CMND 1652. London: HMSO.

Edinburgh Corporation Water Department (1963a). *Report on future sources of supply.* by R.H. Cuthbertson and Associates, Edinburgh (unpublished).

Edinburgh Corporation Water Department (1963b). Papers by consulting engineers in connection with *A preliminary report on amalgamation with the West Lothian Board,* by R.H. Cuthbertson and Associates, Edinburgh (unpublished).

Foster, H.D. and Sewell, W.R.D. (1981). *Water: the emerging crisis in Canada.* Toronto: James Lorimer and Co.

Hanke, S. and Boland, J. (1971). Water requirements or water demands?. *Journal of the American Waterworks Association,* **63**, 677-681.

Hanke, S. and Davis, R.K. (1974). New strategies for water resource planning and management. In: *Priorities in Water Management,* ed. F.M. Leversedge, Victoria, BC: University of Victoria, Department of Geography, Western Geographical Series, **8**, 117-139.

Hammerton, D. (1978). EEc directives on the quality of bathing water and water pollution caused by the discharge of dangerous substances: the River Purification Board viewpoint. *Proceedings of a Symposium on River Pollution Prevention,* The Institute of Water Pollution Control (Scottish Branch). Ingliston, March 1978, 13-28.

Hansard, *Official Report of the Houses of Parliament.* London: HMSO.

Herrington, P. (1982). Water: a consideration of conservation. *Journal of the Royal Society of Arts,* **130**, 332-341.

Hirschleifer, J., *et al.*, (1960). *Water supply: economics, technology, and policy.* Chicago: University of Chicago Press.

Honey, R. (1977). Efficiency with humanity: geographical issues in Scotland's local government reform. *Scottish Geographical Magazine,* **69**, 109-122.

Hunter Committee (1965). *Scottish salmon and trout fisheries. Second Report of the Committee appointed by the Secretary of State for Scotland.* CMND 2691. London: HMSO.

Institute of Sewage Purification (1964). Memorandum on river pollution prevention: urgent need for additional legislation in Scotland. *Journal of the Institute of Sewage Purification,* **63**, 509-512.

Jenkins, R.C. (1973). *Fylde Metering.* Blackpool: Fylde Water Board.

Johnson, J.F. (1971). *Renovated waste water: an alternative source of municipal water supply in the United States.* University of Chicago, Department of Geography, Research Paper **135**.

Johnson, R.W. and Brown, G.M., (1976). *Cleaning up Europe's rivers: economics, management and policies.* London: Praeger.

Kellas, J.G. (1968). *Modern Scotland.* London: Pall Mall Press.

Kelnhofer, G.J. (1968). *Metropolitan planning: some interrelationships.* Georgia Institute of Technology, Water Resources Center, Atlanta, Georgia.

Kneese, A.V. and Bower, B.T. (1968). *Managing water quality: economics, technology and policy.* Baltimore, MD: Johns Hopkins Press.

Lambie, R. (1919). Notes on Lanarkshire (Middle District) Waterworks. Glasgow: Robert Anderson.

McCrone, G. (1969). *Regional policy in Britain.* London: Allen and Unwin.

McIntosh, J. (1966). Drainage and sewage treatment in Caithness. *Journal of the Institute of Sewage Purification,* **65**, 448-451.

MacPherson, H.B. (1970). *Prospects for Metropolitan Water Management.* Cambridge, Mass: Harvard University Press.

Maass, A. (1962). *System Design and the Political Process: The Design of Water Resource Systems.* Cambridge, Mass: Harvard University Press.

Mackie, J.D. and Pryde, G.S. (1935). *Local Government in Scotland.* Dunfermline: Journal Printing Press.

Martin, R.C. (1963). Local adaptation to changing urban needs. In: *Metropolis in transition: local government adaptation to changing urban needs.* ed. R.C. Martin, Washington, DC: US Government Printing Office, 1-12.

Matthew-Fyfe, G. (1950). Pollution in rivers in Scotland. *Transactions of the Royal Sanitary Association of Scotland,* 31-63.

Mitchell, B. and Gardner, J.B. (1983). *River Basin Management: Canadian experiences.* University of Waterloo, Department of Geography Publication Series **5**.

Nicolazo-Crach, J-L. and Le Frou, C. (1978). *Les agences financières de bassin.* Paris: Pierre Johan Et Ses Fils, Editeurs.

Organization for Economic Co-operation and Development (1972). *Water management: gestion de L'eau.* Paris: OECD.

Organization for Economic Co-operation and Development (1976). *Study on economic and policy instruments for water management.* Paris: OECD.

Organization for Economic Co-operation and Development (1977). *Water management policies and instruments.* Paris: OECD.

O'Riordan, J. and O'Riordan, T. (1979). How can citizen input best be utilized by decision-makers? In: *Public participation in environmental decision-making*. ed. B. Sadler, Edmonton: Environmental Council of Alberta, 80-106.

O'Riordan, T. and Sewell, W.R.D. (1981). *Project analysis and policy review*. London: John Wiley and Sons.

Parker, D.J. and Penning-Rowsell, E. (1980). *Water planning in Britain*. London: George Allen and Unwin.

Pierce, J.C. and Doerksen, H.R. (1976). *Water politics and public involvement*. Ann Arbor, Mich: Ann Arbor Science Publishers.

Porter, Elizabeth (1978). *Water management in England and Wales*. Cambridge: Cambridge University Press.

Rees, J. (1974). Water management and pricing policies in England and Wales. In: *Priorities in water management*. ed. F.M. Leversedge, Victoria, BC: University of Victoria, Department of Geography, Western Geographical Series, **8**, 163-192.

River Pollution Commissioners (1872). *Rivers of Scotland. Royal Commission on river pollution*. Parliamentary Papers, **34**.

Reid, W.J. (1956). Drainage and sewage disposal in the county of Aberdeen. *Journal of the Institute of Sewage Purification*, **55**, 169-175.

Royal Commission on Sewage Disposal (1915). *Final Report*, CD 7821. London: HMSO.

Royal Commission on the Distribution of Industrial Population (1940). *Report*. CMD 6153. London: HMSO.

Royal Commission on Environmental Pollution (1972). *Third Report: pollution in some British estuaries and coastal waters*, CMD 5054. London: HMSO.

Scottish Council (Development and Industry) (1952). *Report of the Committee on Local Government in Scotland*. Edinburgh.

Scottish Council (Development and Industry) (1961). *Report of the Committee of Enquiry on the Scottish Economy*. Edinburgh.

Scottish Development Department (1963a). *Central Scotland: a Program for growth and development*. CMND 2188. London: HMSO.

Scottish Development Department (1963b). *Modernisation of local government*. CMND 2067. London: HMSO.

Scottish Development Department (1966). *The Scottish economy*. 1965-1970, CMND 2864. London: HMSO.

Scottish Development Department (1972). *Towards cleaner water*. London: HMSO.

Scottish Development Department (1973). *Measure of plenty*. London: HMSO.

Scottish Development Department (1976). *Towards cleaner water, 1975.* London: HMSO.

Scottish Development Department (1980). *Water in Scotland.* London: HMSO.

Scottish Development Department (1983). *Water pollution control in Scotland. Recent developments.* London: HMSO.

Scottish Water Advisory Committee (1963). *The Water Service in Central Scotland.* Edinburgh: Scottish Development Department.

Scottish Water Advisory Committee (1964). *The Water Service in Ayrshire.* Edinburgh: Scottish Development Department.

Scottish Water Advisory Committee (1966). *The Water Service in Scotland.* CMND 3116. London: HMSO.

Scottish Water Advisory Committee (1972). *The Water Service in Scotland.* London: HMSO.

Schumacker, E.F. (1973). *Small is beautiful.* Abacus, London.

Sewell, W.R.D. (1966). The New York water crisis. *The Journal of Geography,* **65**, 384-389.

Sewell, W.R.D. (1972). Broadening the approach to evaluation in resources management decision-making. *Journal of Environmental Management,* **1**, 33-60.

Sewell, W.R.D. and Barr, L.R. (1978). Evolution in the British institutional framework for water management. *Natural Resources Journal,* **17**, 395-413.

Sewell, W.R.D. and Barr, L.R. (1978). Water administration in England and Wales. *Water Resources Bulletin,* **14(2)**, 337-348.

Sewell, W.R.D. and Coppock, J.T. (1977). *Public participation in planning.* London: John Wiley and Sons.

Sewell, W.R.D. and O'Riordan, (1976). The culture of participation in environmental decision-making. *Natural Resources Journal,* **16**, 1-21.

Sewell, W.R.D. and Phillips, S.D. (1979). Models for the evaluation of public participation programmes. *Natural Resources Journal,* **19**, 337-358.

Sewell, W.R.D. and Rouesche, L. (1974). Peak load pricing and urban water management: Victoria, BC. A case study. *Natural Resources Journal,* July, 383-400.

Sharp, G. (1973). Scotland's water problems treated as second best. *Municipal Engineering,* Supplement, 16 February.

Sherriff, G.J. (1944). A national water policy. *Transactions of the Royal Sanitary Association of Scotland,* 132-162.

Shiell, J.H. (1973). River pollution prevention in Scotland: past, present and future. *Journal of the Institute of Water Pollution Control,* **72**, 261-272.

Slaven, A. (1975). *The development of the west of Scotland: 1750-1960.* London: Routledge, Kegan and Paul.

Tate, D. (1981). River basin development in Canada. In: *Canadian resource policies: problems and prospects,* eds. B. Mitchell and W.R.D. Sewell, London: Methuen, 151-179.

Teclaff, L.A. (1967). *The river basin in history and law.* The Haque: Martinus Nijhoff.

Turing, H.D. (1949). *Fourth report on pollution: Scottish rivers.* London: British Field Sports Society.

United Nations, Department of Economic and Social Affairs (1970). *Integrated river basin development.* New York.

United Nations, Department of Economic and Social Affairs (1973). *Proceedings of an inter-regional seminar on water resources administration.* New Delhi, 22 January - 2 February, 1973. New York.

United Nations, Department of Economic and Social Affairs (1974). *National systems of water administration.* New York.

United Nations, Department of Economic and Social Affairs (1976). *River basin development: policies and planning.* Proceedings of an international seminar, convened in Budapest, Hungary, 16-28 September, 1975. New York.

United States, National Academy of Sciences, National Research Council (1966). *Alternatives in water management.* Report of the Committee on Water, Division of the Earth Sciences. Washington, DC.

United States, National Water Commission (1973). *Water policies for the future.* Washington, DC: US Government Printing Office.

Wengert, Norman (1976). Citizen participation: practice in search of theory. *Natural Resources Journal,* **16**, 23-40.

Wheatley, Royal Commission on Local Government in Scotland (1969). *Report.* CMND 4150. London: HMSO.

Wheatley, Royal Commission on Local Government in Scotland (1968). *Written evidence.* 26 Volumes. London: HMSO.

Wheatley, Royal Commission on Local Government in Scotland (1967, 1968, 1969). *Oral evidence.* 19 Volumes. London: HMSO.

White, G.F. (1961). The choice of use in resource management. *Natural Resources Journal,* **1**, 23-40.

White, G.F. (1964). Choice of adjustment to floods. *University of Chicago. Department of Geography, Research Paper,* **93**.

White, G.F. (1966). Optional flood damage management: retrospect and prospect. In: *Water research,* eds. A.V. Kneese and S.C. Smith, Baltimore, MD: Johns Hopkins Press, 251-270.

White, G.F. (1969). *Strategies of American water management.* Ann Arbor, Mich: University of Michigan Press.

Woodrow, Bruce (1980). Resources and environmental policy-making at the national level: the search for focus. In: *Resources and the environment: policy perspectives for Canada.* ed. O.P. Dwivedi, Toronto, Ontario: McClelland and Stewart.

Index